CLIMATE CHANGE IN INDIA: THREATS, CHALLENGES AND OPPORTUNITIES

Cover photo: Cyclone Tauktae battering the Gateway of India on 17th May 2021

COL C.P. MUTHANNA [RETD]
VICE CHAIR,
KODAGU MODEL FOREST TRUST,
FOUNDER, ENVIRONMENT AND
HEALTH FOUNDATION [INDIA]

Published by

Vij Books India Pvt Ltd
(Publishers, Distributors & Importers)
2/19, Ansari Road
Delhi – 110 002
Phones: 91-11-43596460, 91-11-47340674
Mob: 98110 94883
e-mail: contact@vijpublishing.com
web : www.vijbooks.in

DEDICATED TO

18th

BATTALION

THE MARATHA

LIGHT INFANTRY

CLIMATE CHANGE IN INDIA: THREATS, CHALLENGES AND OPPORTUNITIES

OUR COUNTRY IS MOST VULNERABLE TO THE EFFECTS OF CLIMATE CHANGE AND WE ARE RAPIDLY HEADING TOWARDS A CLIMATE CATASTROPHE.

CLIMATE CHANGE IS NOT AN EVENT THAT WILL OCCUR IN THE DISTANT FUTURE. IT IS ALREADY UPON US AND WE ARE RAPIDLY HEADING TO THE TIPPING POINT. WE DO NOT HAVE THE LUXURY OF TIME. WE MUST ACT NOW!

PRACTICAL AND COST EFFECTIVE SOLUTIONS TO COMBAT CLIMATE CHANGE ARE WELL WITHIN OUR REACH. IF WE MAKE A COLLECTIVE EFFORT WITH A SENSE OF URGENCY,

WE CAN SAVE INDIA FOR OUR FUTURE GENERATIONS AND ALSO LEAD THE WAY FOR THE REST OF THE WORLD.

PREFACE

I have been very fortunate to have served in the Indian Army as an Infantry Officer as this gave me an opportunity to be posted in the Himalayas while my home is thousands of Kilometers to the South in Kodagu, astride the Western Ghats in Karnataka.

This has given me a deep understanding of the Himalayan region and the Western Ghats, both of which are vitally important ecological regions that need to be protected and preserved.

In 1996 I took voluntary retirement from Service with the objective of carrying out active work in the field of environment. Looking back, perhaps it was timely that I returned home, for I was able to comprehend the destruction of the Kodagu landscape which is the Principal Catchment of River Cauvery. While the going was tough in the initial years, the tide has turned and I have now been able to garner considerable support and as a group of NGOs and individuals, we have been able to protect Kodagu from large scale devastation.

PREFACE

(continued)

Meanwhile, it is to be said that you can leave the Himalayas but the Himalayas will never leave you! Concern about global warming and melting glaciers prompted me to work on the HIMEK Alliance proposal for mitigation of Climate Change in the Himalayas and the contiguous Mekong Basin. Persistent efforts resulted in the signing of an MOU between the International Union for Conservation of Nature, Asian Regional Office and the Asian Institute of Technology on the basis of the HIMEK proposal. A HIMEK program has also been initiated at the India Office of the International Union for Conservation of Nature, to carry out pilot projects in some of the Himalayan States in India.

This book does not cover the entire gamut of issues and technologies associated with Climate Change. However, I have highlighted certain very important aspects and I have tried to put forward practical solutions. It is my fervent hope that this publication will help to further sensitize Policy makers in India to the Threats and challenges due to Climate Change and to grasp the opportunities with utmost urgency, before we breach the Tipping Point that is dangerously close.

November 2021

Col CP Muthanna

Athur Village,

Kodagu

ACKNOWLEDGEMENTS

Ms Mallika Sankaran has been of great help in preparing this publication and she has revised and corrected this book innumerable times with utmost patience and care, and I really cannot thank her enough!

I am most grateful to Ms Lily Dechamma for her help in designing this book and I am profoundly thankful to my good friend Jammada Ganesh Aiyanna for printing the first edition of this book free of cost at his printing press in Mysore.

I extend my sincere thanks to Mr Shankar Sharma, Power & Climate Policy Analyst for his most well researched article on the future of the Power Sector in India and to Dr Samir Maithel, Director Greentech Knowledge Solutions Pvt Ltd, for contributing an excellent article on the way ahead for the Brick manufacturing Industry

I am most grateful to Dr CG Kushalappa, Dean, College of Forestry, Ponnampet for his permission to include the very important document prepared by the College on Payment for Ecological Services.

I also thank Mr Somender Singh for contributing the details of his brilliant innovation for internal combustion engines

I wish to express my sincere gratitude to the International Union for Conservation of Nature for the support extended to me in my work on the HIMEK Alliance proposal for mitigation of Climate Change in the Himalayas and the Mekong Region

<h1 align="center">REVIEWS</h1>

Review by Shri B K Singh

The book is authored by Col Muthanna, who has been posted at various places in the Himalayan region during his Army career. He has therefore developed a close understanding of the ecology and environment of Himalayas. After retirement, he has settled in Kodagu (Coorg) and has associated himself with nature conservation in deep Western Ghats areas and promoted a development model that can sustain and leave minimum carbon footprints. His native district Kodagu in the Western Ghats has been experiencing torrential rainfall, flooding and landslides during recent years on account of climate change. Col Muthanna has seen large scale deforestation in Kodagu during the past three decades. This has led to loss of forest habitat and increased human elephant conflict. He has been advocating the restoration of degraded habitats in the region through the COORG WILDLIFE SOCIETY AND KODAGU MODEL FOREST TRUST.

Several scientific papers have been referred in the book. Col Muthanna himself is author/ co-author of some of these. Chapter 2 of the book is a very important paper from Shri Shankar Sharma, who is an expert on energy and a power analyst. He has assessed and suggested several measures for green development and transition from fossil fuel burning to alternate energy from solar, wind and green hydrogen.

In 2009 Col Muthanna initiated a proposal to mitigate climate change in the Himalayas and the contiguous Mekong region. He has proposed a regional cooperation named HIMEK alliance comprising 11 countries with emphasis on drastic reduction of short life climate forcers and forest land restoration. This is an extremely important initiative. Loss of glaciers and depleting water flow in rivers originating from Himalayas would directly impact one-fifth of the human population.

The book has enumerated the impacts of the climate change like sea surges and suggested measures such as relocating vulnerable communities along the coastal regions and providing them with alternate means of livelihood.

The suggestions given in the book would be very useful and important for policy makers, social workers and Governments.

B K Singh,

Former Principal Chief Conservator of Forests (Head of Forest Force) Karnataka

Global warming with the burgeoning anthropogenic greenhouse gas (GHG) emissions (400 parts per million from 280 ppm CO_2 emissions of the pre-industrial era) has been altering the climate, eroding the ecosystem productivity and sustenance of water, affecting the livelihood of people. The anthropogenic activities such as unplanned large-scale developmental activities in the ecologically sensitive regions (Western Ghats, Himalaya), burning of fossil fuels, power generation, agriculture, industrial activities, pollution of water bodies with the mismanagement of liquid and solid wastes are increasing GHG footprint of which 72% constitute CO_2. Land degradation leading to deforestation in the ecologically fragile regions has not only enhanced emissions but also resulted in the substantial erosion of carbon sequestration capabilities. GHG footprint needs to be in balance with the sequestration of carbon to sustain ecosystem functions.

The ecologically fragile Western Ghats has been playing the pivotal role of mitigating carbon footprint with the potential to sequester carbon emission of all southern Indian cities and 1.62% of the total CO_2 emissions from India. India has committed at the Paris Climate Change Agreement to reduce its emissions by 33-35% by 2030, which necessitates the immediate implementation of carbon capture (with afforestation of degraded landscapes with native species, regulations of Land Use, Land Cover changes) and de-carbonization (through large scale implementation of renewable and sustainable energy alternatives) through stringent norms towards (i) protection of ecologically fragile regions, (ii) dis-incentives for continued higher emissions based on 'polluter pays' principle, (iv) adoption of cluster-based decentralized developmental approaches and (iii) incentives for reduced emission. The carbon trading has demonstrated the potential in monetary values across the globe of Indian forests in capturing carbon, and the forest ecosystems in the WG are worth INR 100 billion ($1.4 billion) at $30 per tonne. The carbon credit mechanism and streamlining stakeholder's active participation would dramatically reduce the abuse of forests and encourage farmers to grow trees and convert the land to its next best use.

This book by Col Muthanna presents looming threats of climate change in India which illustrate the challenges such as rising Sea levels, etc. The proposed region-specific land use planning with the climate-resilient mitigation and adaptation strategies would aid in ensuring the sustenance of food and water while ensuring security with minimal instances of disasters. The publication with case studies of Western Ghats and Himalayas, written lucidly, would help in sensitizing decision-makers of the 21st century to evolve strategies toward the prudent management of vital ecosystems.

Dr TV Ramachandra,

Coordinator, Energy & Wetlands Research Group,
Centre for Ecological Studies,
Indian Institute of Science,
Bangalore

Review By Col H S Chauhan

This Book on Climate Change by Muthanna is a work of sheer brilliance and utmost importance. As he has rightly stated, policy makers have to act with a sense of urgency before the 'tipping point' is crossed and we are plunged into Climate Catastrophe. Muthanna has rightly pointed out that mitigation of Climate Change in the Himalayas and preparing for sea level rise should be the most immediate concerns. His initiative in preparing the HIMEK proposal as a regional cooperation of the eleven countries of the Himalayas and the contiguous Mekong basin is a path breaking venture where India must take the lead.

Col Harbhajan Singh Chauhan, SM, VSM, FRGS,

Col Chauhan was a member of the first Trans-Himalayan Army expedition1981-82 that covered a distance of 8000Kms from Arunachal Pradesh to the Karakoram Pass. He subsequently went on to head the Himalayan Mountaineering Institute Darjeeling, Himalayan Mountaineering Institute Manali, and the Indian Mountaineering Foundation at New Delhi.

TABLE OF CONTENTS

Chapter 1

The looming dangers of Climate Change

Climate change and global warming pose the greatest threats and present the biggest challenges ever faced by humanity. The world community is confronted with a bleak future with scarcities of water, food, clean air and land. India will be one of the worst affected countries given our geological and geographical realities. Taken as a whole, the country's extensive coastline and our dependence on the Himalayan Rivers to sustain a large population across the Indo-Gangetic plains in the North put us at high risk from Global Warming and Climate Change.

The commitments made by world leaders including the very important announcements by Our Prime Minister Shri Narendra Modi during the COP 26 Summit at Glasgow could well be a watershed moment for our struggle against climate change. However, this should not lull us into a sense of complacency that all is well, for major concerns remain. It would be better to tread with caution. Two major points for consideration are as follows:

[a] While the Paris agreements goal [similar to the commitments made at COP 26 Summit] was to limit temperature rise to 1.5 degree Celsius, many climate scientists are skeptical and expect the world to warm by at least 3 degree Celsius by the end of the century. There is also the danger that permafrost melt may further dent the calculations on temperature rise due to the massive amounts of methane and Carbon Dioxide released into the atmosphere.

[b] According to the Fourth (2017) National Climate Assessment (NCA) of the United States it is very likely sea level will rise between 30 and 130 cm (1.04.3 feet) in 2100 compared to the year 2000. The report states that a rise of 2.4 m (8 feet) is physically possible under a high emission scenario. This would cause displacement of millions of people in the coastal areas.

Threats and Challenges for India

[a] India's heavy dependence on coal for energy is a matter of serious concern. There is an urgent need to focus on viable alternatives with a greater sense of urgency.

[b] Rising sea levels will force millions of Indians to leave the coastal regions and move inland. Climate refugees from neighboring countries will also flee to India. The intrusion of sea water inland due to sea level rise will render our water sources unfit for human or agricultural consumption due to water salinity. Sea level increases could submerge the infrastructure development projects India is currently investing in, e.g., Mumbai's mega-expensive Sagar Mala project to link the port infrastructures in the country. In the same context, the massive investments planned for tourism infrastructure in the Andaman and the Lakshadweep Islands must also be reconsidered.

[c] The massive river interlinking projects proposed across the country would become unviable. The major concern is regarding interlinking of the Himalayan Rivers. This is due to the alarming increase

in glacier melt and reduced snowfall that would defeat the very purpose of interlinking of these glacier and snow fed rivers.

Reducing India's dependence on coal

During COP 26, India had promised to attain 'Net zero' emissions by 2070.

India is the second largest consumer of coal in the world after China with an annual coal consumption of 966,288,693 tons. Therefore, achieving the net zero targets will be a challenging though not impossible task for the Country. In this context, the following points are for consideration:

[a] **Green Hydrogen holds a lot of promise for India**. Green hydrogen is hydrogen that is produced using renewable energy through electrolysis. This method uses an electrical current to separate hydrogen from oxygen in water. If the electricity needed for electrolysis is generated from renewable sources such as solar or wind, the production of hydrogen in this way emits no greenhouse gases and is termed as Green Hydrogen. Green Hydrogen could form the core of India's Clean Energy mix. Like all fuels, hydrogen when burnt produces energy. But the by-product of burning hydrogen is water, making it the most environmentally friendly fuel. However, the challenge is that there are only a handful of Indian companies currently manufacturing electrolyzers that are required for the production of Green Hydrogen. Moreover, the costs are very high and the cost of green hydrogen production is Rs360-Rs 450 per kg, whereas there will be a need to bring it down to at least Rs 160 per Kg.

Setting up more manufacturing facilities, indigenous production of important components such as electrolyzers and production linked incentives such as the schemes being rolled out by the government for various other sectors will be the most important steps that Indian industry and policymakers need to take to help bring down costs per unit of green hydrogen output.

[b] Wind and Solar Energy

Production of wind and solar energy requires careful planning and production units should not be set up just for the sake of renewable energy. Since wind and solar energy are produced at different places, there will be a need to integrate the two and merge them into the common energy grid. Specific to solar energy, grid connected rooftop solar has enormous potential. It is also known as SPV system. In this system the DC power generated from SPV panel is converted to AC power by using power conditioning unit and is fed to the grid either of 33kV/11 kV three phase lines or of 440/220 Volt three/single phase line depending upon the capacity of the system which is installed at institution or commercial establishment or residential complex and the regulatory framework specified for respective States.

[d] Cooperation between Government and Corporate Sector

Transition from coal based energy to renewable energy in the shortest possible time frame will require close cooperation between the Government agencies and the corporate sector. In this regard, Niti Aayog and the Ministry for Environment, Forests and Climate Change will have to play an important role on the part of the Government whereas the_Federation of Indian Chambers of Commerce & Industry will need to engage closely with the Government on behalf of the corporate sector.

The Government will need to allot substantial funds towards Green energy. There may be a need to shift sizeable funds from existing subsidies for industries in order to provide incentives for maximizing the production of Green energy.

Sagar Mala Project and Tourism projects for the Andamans and Lakshadweep Islands.

There is a need to review proposed development and infrastructure projects that may be rendered redundant or severely impacted due to sea level rise. These include the Sagar Mala project and the proposed mega tourism projects proposed for the Lakshadweep and the Andaman and Nicobar islands. In the case of the Sagar Mala project, infrastructure for the Sagar Mala project would be severely affected by sea level rise and extreme weather events such as cyclones that will become more frequent due to climate change. In the same context there has to be an in depth analysis if the expenditures for the proposed mega tourism projects for the Andamans and the Lakshwadweep islands are warranted. These projects have been discussed in greater detail in chapter2.

River Interlinking Project

A large number of inter river basin transfers have been proposed and they have been categorized as Peninsular component and Himalayan component. The Ministry of Water Resources has also received a number of proposals from State Governments for intra State River linking projects. The river linking projects are expected to cater for existing water shortages to meet the demands of irrigation, domestic and industrial supply and hydropower. One of the main objectives of the project is to divert water from flood prone river basins to areas that are vulnerable to water scarcity and drought conditions. The river linking proposals are in various stages of the planning stage to include Pre-feasibility report, Feasibility report and Detailed Project Report. The Himalayan River linking component includes proposals for river linking projects involving Nepal [Gandak-Ganga, Ghagra-Yamuna, Sarda Yamuna, Kosi-Mecha and Kosi-Ghagra links] and Bhutan [Manas-Santosh-Tista-Ganga link].

The stated objectives of the project appear to be paving the way for water security and freedom from floods across the country. However, there are serious concerns that the river interlinking project will be a huge mistake at an enormous cost of Rupees 5,60,000 crores at 2002 estimates. The river interlinking project will have over 3000 dams and thousands of kilometers of canals. One single Tehri Dam submerged forty villages and displaced the communities from the place that had been there home for generations. What would be the effect of 3000 dams?!

 A few dams in Karnataka across the Harangi, Hemavathi and Kabini rivers have submerged large tracts of forests and cut off traditional migration corridors for wildlife. This is one of the main reasons that today several parts of Karnataka are facing the brunt of human animal conflict, resulting in deaths, injuries and crop loss. What would be the effect of 3000 dams?! I have myself been injured in an elephant attack in November 2020 and I am fortunate to be alive!

River water reaching the sea results in a mix of fresh and salt water that creates an environment in the estuaries that is conducive to fisheries. Blocking fresh water flow into the sea will adversely affect the ecology at the estuaries and the livelihoods of fishermen communities. Water salinity in coastal areas is a problem for agriculture. When fresh water flowing to the sea is blocked due to construction of thousands of dams, more and more areas in the hinterland will lose soil fertility and will become unproductive due to water salinity as the sea water moves further inland and enters the ground water systems. Thoothukudi district in Tamilnadu is a glaring example of agriculture being destroyed by sea water intrusion.

Sea level rise is already eroding our coast lines. When river waters reach the sea, substantial quantity of silt is deposited in the estuaries. This acts as a protection that helps to restrain sea water rise. Therefore, blocking river water from reaching the sea by constructing thousands of dams would be disastrous for the coastal regions of India.

In the light of these concerns, there will be a need for a complete review of the River Interlinking Project. Drought and water scarcity could be addressed through well executed watershed management schemes. It is also vitally important for the Ministry of Water Resources to frame a National Policy for Protection of Catchment areas in the Country, coupled with Payment for Ecological Services to ensure financial security for the communities in these rural areas. Flood water control could be carried out by various nature based solutions. Some dry dams could also be constructed as reservoirs where flood water can be diverted and some intra state river linking could be carried out for the Himalayan rivers systems. All this could be achieved at a fraction of the financial outlay required for the thirty River linking schemes that are now being planned across the Country.

The major concern with regard to the fourteen river interlinking projects planned for the Himalayan Rivers. The Himalayan glaciers are melting rapidly. Studies conducted by the Indian Space Research Organization (ISRO) show that approximately 75 percent of the Himalayan glaciers are retreating at an alarming rate. Many of these glaciers are set to disappear completely or be severely degraded by the end of the century. This could mean that thousands of dams and canal networks constructed for the river interlinking project would be with little or no water. They would be mute testimony of a failed strategy at enormous cost; this would then be the ultimate Himalayan Blunder!

The manner in which India is able to take purposeful steps to mitigate these threats of rising sea levels, eroding coast lines and melting glaciers against the backdrop of a growing population compounded by the influx of climate refugees will act as a model for other countries to take similar measures, and to rise up to the challenge of Climate Change. If our policy makers can make an objective assessment of the situation and take appropriate steps with a sense of urgency, then India can intervene from a position of expertise and action, as a major stake holder in climate change negotiations and discussions at a global level.

On another plane, it is a well known fact that livestock and meat consumption are major drivers of climate change. The worlds demand for meat products result in vast swathes of forests being cut down for grazing, while methane emissions from livestock are a major contributor to climate change. India is culturally and traditionally in a position to encourage reduced meat consumption and vegetarianism and serve as a model for other countries. This could be along the lines of the 'Miss a Meal Campaign' by Shri Lal Bahadur Shastri, that was initiated to overcome the food shortage facing the Country during 1964, when he was the Prime Minister of India. India could lead the way for the global community by declaring one day in a week as a 'no-meat day'.

The solution: a Four-pronged strategy for coordinated Development

1. **Cutting down India's dependence on Coal and rapidly transiting towards Renewable Energy**

2. **Preparing for sea level rise**

3. **Zoning of the Indian landscape to prepare for Climate Change**

4. **Stabilization of Climate Change in the Himalayas and the Mekong Basin**

The solutions to the threat of Climate Change in India will emerge from a true understanding of the causes and an analysis of the effects on the Indian landscape and its people in the years to come. That is the reason for this document and our suggestion of the above four-pronged approach.

The climate crisis is already at our doorstep but if concerned Government agencies take a coordinated approach to mitigate climate change we can minimize the adverse effects. All development and infrastructure projects must take climate and environmental factors into account.

Planning for locating displaced populations from the coastal cities and other coastal regions of India must begin. Development and infrastructure projects must be reviewed in the face of climate change data with a time line of not beyond 2050. Size-able expenditures must be shifted towards building climate resilience (from some of the presently planned projects such as interlinking of rivers).

CHAPTER 2

MOVING AWAY FROM FOSSIL FUELS FOR SUSTAINABILITY

An example discussion through excerpts from the book "A Roadmap to Tamil Nadu's Electricity Demand- Supply by 2050"

By Shri Shankar Sharma

Synopsis: The kind of global climate emergency, as acknowledged by UN Secretary General, UNFCCC, UNEP; as projected by IPCC; and as warned by COP26, have all can be attributed to one common requirement: to minimize the global energy demand, and to meet all our energy needs through a set of renewable energy sources, so as not to transgress the planetary boundaries. A diligent review of our over reliance on conventional technology electricity sources, such as coal, in the context of looming threats of Climate Change should evidently indicate the critical need to move away quickly from fossil fuels in the immediate term, and away from dam based hydro and nuclear power sources also in the longer term. The relevant discussions on the basis of a case study of Tamil Nadu's electricity demand -supply by 2050 can provide a basic framework as to how our country can base its energy transition efforts.

Preface:

From the global perspective of effectively addressing the fast-looming threats of Climate Change, as has been highlighted by the recent COP26 meet in UK, it suffices to say that all possible efforts should be directed in minimizing the global demand for materials and energy. Since the global demand for various forms of energy is also determining the global demand for all kinds of materials, it also suffices to say that the concerted efforts to contain the demand for energy should get the highest priority, as has been the focus of various global summits on climate change in recent years. Within the larger perspective of energy sector, electricity can be expected to occupy a major role within a few decades, as it is increasingly been recognized that the usage of energy in the form of electricity is most convenient and green. It is also techno-economically feasible to project electricity to be the preferred form of energy usage in most of the applications by 2050/60. Efforts are made in the discussion in the paragraphs below to diligently consider the present scenario of Tamil Nadu, and to project a pathway for the state to smoothly transition to a sustainable energy scenario by 2050. The associated though process and due diligence can also be applied to the energy transition scenario of the country.

In the context of India's commitment at COP26 to become net-zero carbon emitter by 2070, all our efforts should focus on this target.

As in the case of any other sector of our economy, in the electricity sector also the choice of appropriate technology for the future must take into account the critical issues such as cost-effective access to resources, global warming considerations, technologies which are already matured and relevant to the state, and the true costs and benefits to the state of such technologies in the long run. A mix of such technologies in electricity generation, transmission, distribution and utilization have to be carefully chosen after due diligence, including effective consultation with stake holders. It may turn out to be that many of the technologies being used today may not be suitable for the year 2050. Though it is difficult to forecast as to which technology may be ideal for a time frame 35 years away, a diligent approach from now onwards will certainly help in minimising the risk of choosing a wrong technology for the future. Whereas there has always been a need to view technologies such as fossil fuel power and nuclear power very carefully in the context of the long-life cycles of these technologies, the fast-looming threats of Climate Change has made it clear that the fossil fuels such as coal and gas must not be a part of our future energy scenario.

India's present energy scenario, as far as energy access to all is concerned, can be said to be far from encouraging from any perspective. Keeping in proper perspective the importance of adequate energy access for human development, the choice of appropriate energy technologies to meet the legitimate energy needs of various sections of the society has become a critical issue for the future.

Factors impacting the choice of electricity generation technology:

Keeping in view the fact that power production technology determines the other associated technologies (such as transmission and distribution) deployed in the electricity sector, adequate deliberations on this aspect of technology becomes very critical.

Comparison of electricity technologies in a matrix form

Technology	Positives	Negatives/Concerns	Ranking & Suitability
Conventional Power Generation Technologies			
Coal	Fairly mature technology; all weather and reliable source of electricity; suitable for base loads; has been main source for many decades	Pressure on land, water and minerals; fast depleting fuel source; social and environmental concerns; ever increasing costs and supply risks; global warming concerns; disposal of ash; pollution of land, water and air.	Lowest Rank; Least Suitable to India
Natural Gas	Much less pollutants and GHG concerns; suitable for peak loading because of its ability to start and stop quickly; can be considered as a link fuel between coal-based power era and renewable energy era.	No large-scale domestic reserves; huge import dependence; fracking has huge global warming concerns; needs a lot of fresh water; concerns of pollution of land and water	Lower Rank; Not much potential in India
Dam based hydro	Clean source of electricity; operational costs are very small; best suited for quick start, stop or varying loads; can be a part of multipurpose dams; considered a renewable energy source	Many social and environmental concerns; Methane as highly potent GHG; location specific and largely away from load centres; mostly in hilly terrains; need long transmission lines; much larger construction time; vastly mpacted by the vagaries of rainfall; likely to be impacted by the climate change	Low Rank; Not much potential in the long run

Mini/micro hydro	Has least costs on social and environmental grounds; highly suitable for hilly terrains; can implement photovoltaic solar energy systems during winter/ rainy seasons	Not highly suitable for grid-based operations; due to absence of vast storage capacity, the reliability can be low	Good Rank; Not much potential
Nuclear	No GHGs during its operation; Low operational costs; highly suited for base load operations	Capital costs and risks are very high; huge threat of accidents and radiation; spent fuels have remained a serious issue and can have dangerous levels of radiation for thousands of years; not enjoyed popular support	Lowest Rank; Least Suitable to India
New and Renewable Energy Technologies			
Solar	Clean, green and renewable; most suitable at consumer's premises; no pressure on land and water; minimum pollution loading; fast declining prices; gaining lot of support; low operational costs	Per MW Capital costs are considered high; suitable only during sun-shine hours; intermittency is a serious issue; may need back up or storage in many applications	Highest Rank; Highly Suitable
Wind	Clean, green and renewable; minimum pollution loading; fast declining prices; gaining lot of support; low operational costs	Intermittency is a serious issue; may need back up or storage in many applications; location specific; can leave considerable foot print on environment	Highest Rank; Highly Suitable
Bio-mass	Clean, green and renewable; highly suitable where large quantities of agricultural and plant wastes are available; suitable to rural areas where electricity has not reached; bio-energy has been in use in India for thousands of years; low costs and environmental foot print.	Unregulated usage can lead to food security issues; bio-electricity production is relatively new, and may not be seen as a mature technology	High Rank; Highly Suitable
Geo-thermal	Clean, green and renewable; cost effective and reliable; has been in usage in New Zealand an Iceland for decades	Location specific; historically limited to areas near tectonic plate boundaries	High Rank Suitability to be studied
Ocean energy	Has huge potential for the long Indian coast; can have very low operational costs	Not a mature and widely used technology; can have impact on marine creatures; needs more of R&D spending.	High Rank Suitability to be studied.
T & D system			
Integrated Network	Has come to be known as essential; needed in case of large size power plants; has been very popular and	Complexity and risk of failures growing; has not met the needs of rural areas in India; has social and environmental footprint;	High Rank; Considered essential; but suitable

	widely used.	increasing costs; land acquisition is an issue; prone for natural calamities such as flood and storms.	modifications are needed,
Micro/ Smart Grid	Most suitable for smaller / rural communities; provides local control; low capital and operational costs; highly suitable for distributed REs; has nil or very low social and environmental footprint; much easier to manage.	Emerging technology; will need very advanced protection and communication technologies; stake holders have to exercise much higher operational discipline.	Highest Rank Essential and suitable.

A high-level analysis of this matrix format itself should be able to provide clear indication to our policy makers in their choice of suitable power technologies for the future. Even without considering the need to keep Climate Change in focus, it should become obvious that our energy scenario for the future should have no place for fossil fuels such as coal and natural gas.

Climate Change considerations

Few years ago, an article in 'The Guardian' had state: "Three decades from now the world is going to be a very different place. How it looks will depend on actions we take today. We have big decisions to make and little time to make them if we are to provide stability and greater prosperity to the world's growing population. Top of the priority list is climate change." This statement sums up the challenge before the humanity. The evolving threats associated with the Climate Change to the human civilization has been accepted as so critical that no planning decisions and policies for the future seem to be acceptable without focusing on the core issue of Climate Change. Electricity sector, being the major contributor to the global warming phenomenon and also since its infrastructure can be highly vulnerable to the vagaries of the climate change, can be no exception in this regard. Subsequent to 26 meetings of COP under the aegis of UNFCCC, this scenario should become evidently obvious.

IPCC Reports and other global reports on Climate Change

The UN Intergovernmental Panel on Climate Change (IPCC) insists that unless global warming is addressed adequately and early, the planet promises to suffer all manners of evil. Because of coastal flooding and storm surges, urban populations are especially susceptible to the risk of death, injury, and disrupted livelihoods. IPCC in its five Assessment Reports (AR1 to AR5) has not left any doubt regarding the anthropogenic reasons for the global warming, while emphasizing that the burning of fossil fuels is a major contributor of Green House Gas (GHG) emissions.

The fifth assessment report of IPCC (AR5) had indicated that emission of the greenhouse gases must fall by 2050 by 50-85% globally compared to the emissions of the year 2000, and that the global emissions must peak well before the year 2020, with a substantial decline after that. As per this report "Emissions from deforestation are very significant – they are estimated to represent more than 18% of global emissions"; "Curbing deforestation is a highly cost-effective way of reducing greenhouse gas emissions." As per the Ministry of Environment, Forests and Climate Change (MoEF&CC), about 33% of all the coal reserve in India is below thick forests. Planning for additional coal power plants will mean opening of more of coal mines which will lead to destruction of these thick forests.

The huge contribution of GHGs from energy/electricity sector is clear from IPCC reports, where electricity and heat production is shown as responsible for 25% of the global GHG emissions. In the case of India, the energy sector is associated with about 58% of all GHGs in the country as per the statement by MoEF&CC to the Parliament. MoEF&CC report of 2010 also says that the electricity sector contributed about 38% of all GHG emissions in the country during 2007. This indicates the major role of the electricity sector in GHG emission efforts for the country.
{http://timesofindia.indiatimes.com/home/environment/global-warming/Energy-sector-accountsfor-58-of-greenhouse-gas-emissions Govt/articleshow/50275517.cms}

Within the energy sector the electricity sector has come to be knows as a major contributor of GHGs. The inevitable linkage between global electricity usage and CO2 emission is no more an issue of uncertainty. The higher usage of electricity is clearly associated with higher emission of CO2.

Global Warming and India

Being a tropical country, India is projected to be one of the worst affected countries, if global warming phenomenon is allowed to cross the limits. Some of the projected impacts can be listed as below.

> • Snow fed Himalayan rivers Ganga, Yamuna, Brahmaputra etc.; are seriously threatened; initially there could be heavy floods, and then only seasonal rivers;
> • Low lying coastal areas on east coast may face submergence;
> • There could be unpredictable weather; storms and hurricanes will become more frequent; so will heat waves and droughts.
> • Threats ranging from cloudbursts, avalanches, landslides, to glacial lake outburst floods will increase;
> • Vastly decreased food production; increased tropical diseases;
> • Increased rain fall in some areas, and grossly insufficient rainfalls in other areas; a definitive increase in drought prone areas;
> • Water scarcity and possible water wars between communities within the country.
> • Thermal power plants demand large quantities of fresh water. Studies have determined that in US about 50% of fresh water consumption is for thermal power production. It is also well known that it takes a great deal of water to supply energy and a great deal of energy to supply water associated with activities such as pumping, transporting, treatment and desalination.

Climate Risk and Adaptation in Electric Power Sector

Whereas the power sector is closely associated with the causes for global warming, the phenomenon of Climate Change itself can impact the electric power sector in many ways. Asian Development Bank's (ADB) Year 2012 report "Climate Risk and Adaptation in the Electric Power Sector" has discussed such issues as applicable to Asian countries. It is well known that the electric power investment decisions have long lead times and long-lasting effects, as power plants and grids often last for 40 years or more. This necessitates the need to assess the possible impacts of climate change on such infrastructure, to identify the nature and effects of possible adaptation options, and to assess the technical and economic viability of these options. The report has stated that the power sector is vulnerable to projected climate changes, including the following:

- Increases in water temperature are likely to reduce electricity generation efficiency, especially where water availability is also affected. Such a scenario has impact on thermal power generation.
- Increases in air temperature will reduce electricity generation efficiency and output as well as increase customers' cooling demands, stressing the capacity of generation and grid networks.
- Changes in precipitation patterns and surface water discharges, as well as an increasing frequency and/or intensity of droughts, may adversely impact hydropower generation and reduce water availability for cooling purposes to thermal and nuclear power plants.
- Extreme weather events, such as stronger and/ or more frequent storms, can reduce the supply and potentially the quality of fuel (coal, oil, gas), reduce the input of energy (e.g., water, wind, sun, biomass), damage generation and grid infrastructure, reduce output, and affect security of supply.
- Rapid changes in cloud cover or wind speed (which may occur even in the absence of climate change) can affect the stability of those grids with a sizeable input of renewable energy, and longer term changes in these and precipitation patterns can affect the viability of a range of renewable energy systems.
- Sea level rise can affect energy infrastructure in general and limit areas appropriate for the location of power plants and grids. This scenario may impact the power plants located or planned close to coastal areas in TN.

T&D grids can be highly sensitive to high ambient temperature (increased electrical resistance) and storm damage. Energy companies are more often cited as part of the problem of climate change, generating the lion's share of the world's greenhouse gas emissions, amounting to around 40% of the total. But they will also suffer as global warming picks up pace, as generators – from nuclear reactors to coalfired power plants – feel the brunt of the weather changes.

Even though natural gas (LNG) is considered as a lesser evil as compared to coal and diesel, and its usage may be seen as a link energy source for countries like India, before REs replace the fossil fuels, the huge potency of Methane as a GHG (ranging from 86 to 105 times more than that of CO2 as per reports) should not be overlooked. When the estimates for methane leakage based on actual observations in the upstream processes and transportation are objectively concerned, and when the natural gas is transported over long distances, its global warming potential is reported to be quite high; not much less than that of coal. Hence the natural gas option for India and Tamil Nadu cannot be good options either. A US govt. report "Life Cycle Greenhouse Gas Perspective on Exporting Liquefied Natural Gas from the United States" has focused on this issue.

(http://energy.gov/sites/prod/files/2014/05/f16/Life%20Cycle%20GHG%20Perspective%20Report.pdf)

Recommendations of 'expert group on low carbon strategies for inclusive growth'

The 'expert group on low carbon strategies for inclusive growth' which was set up under the erstwhile Planning Commission of India to develop a strategy for India's 12th Five Year Plan had released its final report in May 2014. Its main findings were:

- India will have to invest $834 billion in the two decades ending 2030 to reduce its emission intensity to gross domestic product (GDP) by 42 per cent over 2007 levels
- The massive change in the energy mix by 2030 will result in lower annual demand of coal at 1,278 million tonnes from an estimated 1,568 million tonnes.
- Under the low carbon energy mix, the installed capacities of wind and solar power would have to be increased to 118 GW and 110 GW, respectively, by 2030.
- The huge investments needed in low carbon strategy would have little impact on economic growth.
- The report also highlights the importance of more efficient coal power plants in the future and the use of renewable energy resources.
- It suggested that the aim should be that at least one third of power generation by 2030 be fossil-fuel free.

Compliance with various Acts of our Parliament

When we view our own accumulated experience since independence, of the power sector from the perspective of various Acts of our Parliament the gap between such mandates and compliance may become obvious.

While it has been recognised that it is impossible to predict the impacts of climate change with great degree of accuracy, there can be no doubt that there will be increased warming and impacts throughout the century across the globe. For example, sea-level rise will affect hundreds of millions of people regardless of how we control greenhouse gases now. Hence, we have to adapt to that likelihood. Hence, taking drastic action now to lower emissions would give us a better chance of avoiding the worst climate effects. Cutting emissions would also push back the effects of climate change by several decades, giving us more time to adapt.

The conventional power plants and the associated infrastructural elements such as coal mines, coal transportation network, coal storage yards, ash ponds, hydro reservoirs, transmission lines, nuclear ore mines, spent fuel storage facilities etc., have been throwing up many serious concerns (social, economic, environmental and inter-generational issues) to our communities since independence which cannot be ignored anymore, especially in the context of global warming. Forced displacement of the project affected families is a common but credible threat to our communities because of each of the conventional power generation technologies. Loss of livelihood; denial of access to stretches of forests, rivers and oceans; inadequate or nil compensation; destruction of habitats etc. have impacted the lives of millions of people from such projects since independence.

As per the sections 48 (a) and 51 (a) (g) of our Constitution it is the duty of the State and every citizen to make honest efforts to protect and improve our environment by protecting and improving rivers, lakes, forests and living beings. When we also objectively consider our own accumulated experience of conventional power plants since independence, our inability/indifference to comply with the letter and spirit of various Acts of our Parliament namely the Environment Protection Act, Forest Conservation Act, and Wildlife Protection Act, and national forest policy will become glaring. These Acts have all emphasized and / or mandated the need to preserve a healthy environment, adequate quality and extent of forests, and to protect and enhance bio diversity in the country. Since the electricity sector has been known to have considerable impact on various aspects of nature, the relevant provisions under these Acts also need to be kept in our focus all along the planning and implementation of the future electricity infrastructure for the state.

Like all other infrastructures, electricity infrastructure too must be designed, redesigned, and reviewed continuously so as to meet the enormous challenges of the fast-evolving global warming phenomenon. In summary, it should be mentioned that an unending growth to electricity demand, and hence to economic growth, is not in the interest of the human kind in the long run. All possible measures should be taken to protect our natural resources from the ravages of a poorly managed electricity industry.

Renewable energy potential in TN

Whereas TN already has about 36 % of its installed generating capacity in the form of renewable energy sources, it has vastly more potential. In a 2012 study done by the World Institute for Sustainable Energy (WISE) with support from Shakti Sustainable Energy Foundation on the potentials for renewable energy for the state, the results showed the following: onshore wind potential, together with grid-tied solar PV and solar CSP, can contribute about 535,059 MW as against a total estimated potential of 682,800 MW. In addition, offshore wind energy potential is about 127,428 MW, a majority of which comprises very high resource quality areas with net capacity utilization factors of over 30 percent.

[Reference: Renewable Energy Potential for Tamil Nadu; *TN State action Plan on Climate Change, 2013]*

An overview of the RE potential and demand by 2050

The vast potential of RE in TN can become obvious when we consider the potential of roof top SPVs alone. Assuming about 1.8 crore houses in the state (at 25% of the population of 7.2 crores in January 2014), and that about 50% of these houses can be considered to be economically and structurally suitable to house roof top SPVs of capacity 5 kW each (@1 kW per 100 sq. ft) a potential of about 45,000 MW can be projected in the residential sector alone. If we also consider the huge roof top surfaces available in educational institutions, offices, commercial and industrial sectors, warehouses, hospitals, hotels, and other buildings the potential is huge (running into few millions of MW) thereby minimizing the need for other kinds of REs. The advantage of such roof top SPVs is that they drastically reduce the effective T&D losses, as they produce power where it is also consumed, and they also eliminate the need for land acquisition.

The total RE potential of about 740,000 MW as per *TN State action Plan on Climate Change, 2013* is huge as compared to the projected power requirement of TN in the foreseeable future. The projected electricity demand by Year 2050 (215,420 MW of peak power and 1,492,465 MU of annual electrical energy at 8% CAGR, and 109,900 MW/ 761,480 MU at 6% CAGR) in a business-as-usual scenario can be considered for further analysis. It seems safe to assume that electricity demand growth at 8% CAGR for next 35 years is impossible. Additionally, knowing well the implications of such a growth it will also be safe to assume all out efforts will be made to reduce such growth rate. Hence a conservative growth rate of 6% CAGR can be assumed through to 2050. Considering the huge potential through conservation measures feasible in the state, it may also be assumed that the projected demand for Year 2050 at 6% CAGR can be further reduced by about 25%, which may mean a demand of about 85,000 MW of power and 500,000 MU (or 500 Billion Units) of annual energy. The total RE potential of 742,561 MW in the state at an average of 20% utilisation factor can yield about 1,300 Billion Units of energy. It becomes clear that the annual RE potential is about 2.3 times the projected energy requirement by 2050. A unique characteristic of Solar PVs is that they are generally modular in nature, and the capacity can be easily increased by adding PV panels at the existing site itself. Hence, the total potential of roof top SPVs can be many times more than that indicated earlier. Hence, meeting an increased annual energy demand should not be an issue in the case of those installations having roof top SPVs. This unique feature increases the energy potential of SPVs enormously in India's efforts in rapid energy transition.

The Modi government's climate change and development plan

Melbourne Sustainable Society Institute, Australia has prepared a series of briefing papers that examine the climate change policies of the countries key to a global agreement at the United Nations Framework Convention on Climate Change (UNFCCC) negotiations in Paris. The briefing paper on India discusses the transition in India's cities and urban transport sector – towards clean eco-friendly urban spaces and infrastructure. The paper has concluded that such a transition has the potential to produce strong social (health), economic (jobs and investment) and environmental (lower GHG emissions) dividends for India.

The share of renewable energy discussed by Indian government institutions, both in absolute terms and relative to other sources of electricity, has become increasingly ambitious over the years, ranging from:

- 2009: National Action Plan on Climate Change (NAPCC) target of 15 percent of total electricity consumption from RE by 2020

- 2010: National Solar Mission (NSM) target of 20 GW of solar by 2022

- 2013: PGCIL Desert Power 2050 estimate of 458 GW of wind and solar by 2050

- 2014: Discussions pertaining to a solar target of 100 GW by 2027

- 2014: Consultation pertaining to a National Wind Energy Mission in which a wind target of approximately 150 GW by 2027 is being discussed

- 2014: Estimate by the NITI Aayog's "heroic effort" scenario of 410 GW of wind and 420 GW of solar by 2047

- 2021: India's commitment to become net zero by 2070 and to have 500 GW of RE power capacity by 2030

(Ref: "Report on India's RE Roadmap -2030" MNRE, Feb. 2015)

Such a plan indicates a growing confidence on the potential of RE in India, and the government's ambition to move towards a RE dependent power system.

India's own experience

The usage of non-conventional energy sources (OR new and renewable energy sources) is not an entirely new phenomenon to India. Our ancestors have been using the sun's energy and wind energy in many ways for thousands of years. In recent years the usage of off-grid REs has taken off in a big way, mostly due to private initiative. The ministry of non-conventional energy has also taken many initiatives to popularize the usage of REs. India's INDC to UNFCCC in Nov. 2021 (COP26) indicates its own confidence in RE technology with a target of 500 GW of RE by 2030, and 40% of all power capacity through non-fossil fuels. This INDC can be seen as a very strong statement in India's commitment to reduce its reliance on fossil fuels. TN's power sector plans have to keep this national target at their focus, especially since it has no fossil fuel reserve of its own except for Lignite.

International Solar Alliance (ISA) is an initiative led by India with membership from the solar resource rich countries lying fully or partially between the Tropic of Cancer and the Tropic of Capricorn. About 120 prospective member countries have been identified for this purpose, and the alliance is expected to transform the way solar power is harnessed across the world for the benefit of all these member countries, especially for the disadvantaged communities. The much-required cooperation and coordination at the global arena is proposed to be filled by ISA, whose member countries are keen to transform their solar resource wealth into improved lives for their people through application of solar technologies. These countries can potentially harness solar energy in a cost-effective manner, if a concerted and coordinated effort is made to share experience from other similar countries and concentrate on finding solutions which are designed to be locally appropriate for difficult conditions, while still remaining affordable. The overarching objective of ISA is to create a collaborative platform for increased deployment of solar energy technologies to enhance energy security & sustainable development; improve access to energy and opportunities for better livelihoods in rural and remote areas and to increase the standard of living.

Rajasthan experience in Solar powered IP sets

Two survey reports (i) "Solar Irrigation Pumps: Farmers' Experience and State Policy in Rajasthan" in Economic & Political Weekly of March 8, 2014; and (ii) "Solar Irrigation Pumps: The Rajasthan Experience" by Nidhi Prabha Tewari (www.iwmi.org/iwmi-tata/apm2012) have provided detailed information on the experience in Rajasthan where the Government of Rajasthan (GoR) had launched a Rs. 515 crore scheme in 2011 to provide subsidized solar irrigation systems to 10,000 farmers in the state over three years. In 16 districts of the state, 1,675 farmers got solar pumps of 2,200 or 3,000 Wp at a subsidy of 86% in the first year of the scheme (2011-12). Most of these pumps were installed in farmers' fields in the summer March-June) of 2012. In the second year, the state government plans to install an additional 4,500 solar pumps in all 33 districts of Rajasthan. These two reports have come to the conclusion that the experience of the farmers in this regard has been mostly encouraging. The first report says: "Solar pumps are convenient to use, require minimal attendance and have few maintenance problems. Each 3,000 Wp system saves its owner Rs. 45-65,000 worth of diesel, besides increasing land and water productivity and crop quality. It also saves him labour and exposure to noise and air pollution. All owners we talked to were very happy with their PV pumps. They hoped to recover their share of the system's cost in less than two years. Each one of them thought other farmers should get solar pumps."

A Rajasthan government document claims that an investment of Rs.700 crore can save the installation costs of 70,000 new agriculture electricity connections to farmers.

Study by WISE, 2012 – "Action Plan for Comprehensive Renewable Energy Development in Tamil Nadu":

This study by World Institute of Sustainable Energy, Pune, in 2012 has dealt with the reassessed RE potential in TN, the simulation of TN system with aggressive penetration of RE in 12th and 13th Plan periods, and has made high level estimates of costs and benefits to the state. Its key findings are:

- There are major risks involved in the 'business-as-usual' approach associated with coal dependent power planning in the state. The focus of energy planning has to essentially shift from a mere matching of supply-demand, to an approach guided by long-term energy security. This is where renewables can add value and contribute to long-term energy security for the state.

- The use of modern methodologies including Geographic Information System (GIS), has shown that the total (reassessed) RE potential for Tamil Nadu is over 720,000 MW (including grid-connected and off-grid power). The constrained potential is about 530,149 MW.

- The study shows that an aggressive RE integration scenario is technically feasible, provided operational philosophies are modified and evacuation infrastructure is built to integrate RE.

- Under the aggressive RE integration scenario, RE evacuation studies for the 12th plan showed that in order to evacuate 23,613 MW of RE, (Proposed 16,660 + Existing 6,953 MW), additional 51 substations of 230 kV and 60 substations of 110 kV are required. In order to evacuate 43,423 MW of RE (Proposed 36,470 + Existing 6,953 MW) in the 13th plan, additional 74 substations of 230 kV and 61 substations of 110 kV are required.

However, for both the plan periods, large increases in power flows are expected due to injection of solar power into the grid.

- The estimated cost for the total evacuation infrastructure over the 12th and 13th five-year plans to support aggressive RE capacity addition is Rs. 11,025 crore, and is very small as compared to the total funds planned for transmission and distribution capacity augmentation under the Vision 2023 document (Rs.200,000 crore).

• Aggressive RE integration scenario is projected to provide monitory and other commercial benefits to the state. The study suggests that contrary to general perceptions, commercial implications of RE are not very high as compared to risks involved in other purchase options such as that of conventional power projects, whose price is variable and increases due to delays, cancellations, etc.

Study by WISE and WWF, 2013 - The Energy Report – Kerala: 100% Renewable Energy

This study report of 2014 by World Institute of Sustainable Energy, Pune and WWF has attempted to model the energy requirement (across power, transport, agriculture, industry, domestic and commercial sectors) of Kerala up to 2050 in order to assess the feasibility of meeting 100 percent of the state's energy demand with renewable sources. Its main findings are:

• The central finding of the study is that Kerala can meet over 95 per cent of its energy demand with renewable sources by 2050.

• 100 per cent electricity requirements for the state can be met with RE.

• The main resources are onshore and offshore wind, grid-tied and decentralized/off-grid solar, large and small hydro, biomass and wave energy.

• Whereas in a business as usual (BAU) scenario the total energy demand of the state by 2050 would be about 5.5 times that of the total energy in 2011. In stark comparison in a curtailed demand scenario, which was assessed without compromising economic output (but by aggressive interventions in energy efficiency, energy conservation and carrier substitution), the total energy demand by 2050 was only 2.2 times that of the total energy in 2011.

• A BAU growth cannot be sustained indefinitely even with all fossil and renewable sources unless we are able to decouple economic growth (GDP) from energy resource use.

Financing the Renewable Energy

The direct and indirect subsidies to fossil fuel industry all these years run to many Trillions of Dollars at the global level. Compared to it the money spent so far to encourage REs can be said to be tiny. International Energy Agency has shown, in the 37 countries it analysed, oil, gas and coal received $409 Billion in 2010 compared with $66 Billion for renewable energy. Keeping in view the huge costs to the society from the implications of global warming, the society must make all possible efforts to divert such vast subsidies from fossil fuels for developing and deploying the REs. So, fundamentally it should cost no additional moneys to the society to switch over to an era of complete reliance on REs.

When we consider the geographic, climatic and resource strengths/ constraints of TN, and the global warming implications between now and 2050, the nature of the electricity infrastructure for the future should become fairly clear. It looks eminently credible to suggest that it should be based on a very large number of small size, distributed renewable energy sources supported by modern, highly efficient and smart T&D infrastructure, including micro/smart grids. All the discussions in the earlier sections should give adequate confidence for the state to move resolutely towards 100% renewable energy sources largely dependent on distributed REs, micro and smart grids.

TN's case is similar to that of the other Indian states in most aspects. The country's inability to provide adequate quantity of clean, affordable and sustainable electricity is a major constraint in achieving energy security. The present centralized model of power generation, transmission and distribution is growing more and more costly to maintain and, at the same time, restricts the flexibility required to meet growing energy demands, and to satisfactorily meet the expectations of every consumer. The exhaustible nature of the fossil fuels, and their contributions to human health issues the and global warming have persuaded/ forced the global leaders to agree to do all that is needed to eliminate /neutralise the GHG emissions from the fossil fuels. TN does not possess any fossil fuel reserve except for a limited amount of Lignite, which is considered as the worst form of fossil fuel. Since TN is also constrained in fresh water resources, which will be needed for coal power generation, the state has no option other than to look for ways to minimise/ eliminate its dependence on fossil fuel-based electricity.

India and TN are in a critical need to adopt a massively driven decentralized economic development model in order to more readily take advantage of the abundantly available renewable energy sources like solar, wind, hydropower, biomass, ocean energy, geothermal and hydrogen energy, and fuel cells. India (so is TN) is blessed with an abundance of these resources, yet it spends millions of rupees to import oil, coal, and natural gas resulting in enormous amounts of renewable energy being unused/ wasted. To that end, renewable resources are the most attractive investment because they will also provide long-term economic growth for India. To secure its energy future, India /TN urgently needs to design/implement innovative policies and mechanisms that promote increased use of abundant, sustainable, renewable resources. All of its future energy demand could be met by an optimal combination of utility-scale and rooftop PV, concentrated solar power, onshore and offshore wind, bio-mass, geothermal and conventional hydropower. Ocean energy, though can be huge because of the long coast line, is not a mature technology yet, but can be expected to be developed further.

The future power sector scenario would require building a large no. of solar power systems and wind farms, hybrid solar-natural gas plants, solar thermal storage, bio-energy and advanced battery-based grid energy storage systems. Investment in these technologies would create millions of new jobs and humongous economic stimulus, including vast employment potential across the state, especially rural area, if all indirect (ripple) effects are included. Other major changes involve use of electric vehicles and the development of enhanced Smart Grids. Making the transition to 100% renewable energy is both techno-economically feasible, but requires strong political support. Most importantly, TN (and the country as a whole) has no other option but to review the past and present policies, and adopt a sustainable way of meeting electricity demand of its population as discussed in these pages. The societal level principles and the approach to overall welfare issues taken in this study report can be applied to all other states of the Union also with suitable modifications depending on the local geography, climatic conditions, consumer profiles and developmental pathway.

Road Map for Tamil Nadu – 2050

When we consider the geographic, climatic and resource strengths/ constraints of the state, and the global warming implications between now and 2050, the nature of the electricity infrastructure for the future should become fairly clear. It looks eminently credible to suggest that it should be based on a very large number of small size, distributed renewable energy sources supported by modern, highly efficient and smart T&D infrastructure, including micro/smart grids. All the discussions in the earlier sections should give adequate confidence for the state to move resolutely towards 100% renewable energy sources largely dependent on distributed REs, micro and smart grids.

Potential for electricity demand reduction measures

Measures such as efficiency improvement, DSM, energy conservation and effective usage of solar powered appliances have the potential to reduce the demand on the existing integrated power network by a huge margin, as indicated in the table below. Effective implementation of these measures is critical to minimize the effective demand for electricity, and hence also to minimize the material requirement and other associated concerns even in a 100% RE scenario.

High level estimate of potential for additional sources of electricity for TN

Source of virtual additional power OR savings	Estimated Potential for savings	Reference
1. Energy Conservation Potentials for various Sectors in Tamil Nadu	18% (of state's total consumption)	Electrical Inspectorate, Tamil Nadu, 2011 (Table 14 in the book)
2. T&D loss reduction	14% of total (reduction from 19% to 5%)	Tables 5 and 6 in the book
3. Agricultural sector	8 - 10% (of state's total consumption)	Through shifting IP loads to solar power; certain savings through conservation already taken into account in item 1
4. Domestic sector	10 - 15% (of state's total consumption)	Through shifting loads to solar power; certain savings through conservation already taken into account in item
5. Commercial sector	3 - 5% (of state's total consumption)	Through shifting to solar power of lighting and other smaller loads; certain savings through conservation already taken into account in item 1
6. Industrial sector	3 - 5% (of state's total consumption)	Through shifting of smaller loads and lighting loads to solar power; certain savings through conservation already taken into account in item 1

7. Municipal Water works and Street Lighting	1% (of state's total consumption)	Through shifting of all lighting loads to solar power; certain savings through conservation already taken into account in item 1
Estimated total DSM potential	60 - 70% (of state's total consumption at present)	Through efficiency, DSM, conservation and partial shifting to solar power

Assuming that many of the loads, can be shifted to off-grid solar power mode, the demand on the integrated grid can only be about 60-70% of the current demand. This scenario can make the present scenario in the state surplus by a considerable margin.

An important consideration in demand and supply of electricity by 2050 could be that any limitation in the supply of electricity will be likely to be the limitations of the T&D elements than on the electricity generating capacity, because total RE capacity can be scaled up to very large levels (in-situ scenario) unlike conventional energy sources. When REs are effectively deployed at consumer's premises, the constraint of power generating capacity may theoretically vanish. In this context any additional demand for electricity in a scenario of 100 % RE and a large number of micro/smart grids connected to each other through sensitive protection and communication systems shall theoretically pose no problem.

Focus on specific sectors of the economy

In the case of agricultural pumping needs (accounting to about 20% of the total electricity consumption), while it is feasible to meet all its electricity needs by solar-powered pumping technology, which is ideally suited for our farmers who need such pumping during day time and during non-rainy days, such usage can save as much as 25 to 30 percentage of total grid electricity consumed in the state if we also take into account the T&D loss savings. Similarly, the vast quantities of the municipal water pumping loads, can be and should be shifted to solar power and during daytime.

Domestic electricity needs, which is about 27% of the total annual electricity consumption, can largely be met by roof top SPVs resulting in the grid demand reduction by as much as 30% at the state level. Lighting needs of commercial and public places, which together account for about 15%, also can be largely met through SPV systems of suitable technologies. Such efficient lighting systems dependent on solar power are reported to be satisfactorily performing in many cities in the state, and efforts should be made to extend to all areas in the state. The hot water/ steam needs of restaurants, hostels, hotels, hospitals, schools, religious institutions such as temples where community cooking is undertaken etc. can and should be shifted to solar heating systems during day time, assisted with storage systems where necessary.

Industries too can meet almost all their lighting needs, and electricity needs of many of the lighter applications such as small size motive power, heating, cooling, steam generation, process heating, cold storage, computing, control and instrumentation etc. through SPVs/ solar heating systems located on their roofs. There are many mature technologies available in the market to shift many of the present grid electricity loads to solar powered systems. Adequate efforts and investments in this context in industrial and commercial sectors can reduce the electricity demand on the integrated grid by considerable margin.

'SUN FOCUS' is a magazine periodically issued by MNRE, which focuses on various applications on concentrated solar power such as concentrate solar thermal (CST), and the related activities in India. This publication provides an indication of the vast potential identified by MNRE with solar power in the country. It has listed a number of heating and cooling applications, which can make use of solar power in industrial segments such as Pharma, Textiles, Food-processing and Chemicals. As mentioned in this magazine about 170 CST systems have been already installed in the country and thirty more are coming up. Optimal use of the associated technologies can assist in considerable reduction in grid electricity demand, and also can shift many applications dependent on other fossil fuels and bio-mass to solar power. Steam is required in many industrial processes, and so far, this heat is generally generated by fossil fuels such as grid electricity, gas or furnace oil or diesel. Concentrating solar technology can not completely replace these fuels, but in sunny regions they can also augment the existing process to reduce the fossil fuel consumption. The state has to put all possible efforts to optimally harness such technologies supported by REs to achieve a sustainable energy sector by 2050.

On objective consideration of myriad problems with the existing electricity infrastructure, with large size centralised conventional power plants at the centre, and the network of transmission lines, substations, distribution lines and transformers etc., it may seem obvious that the characteristics of future electricity infrastructure has to be vastly different to what we see today. It is reasonable to assume that in view of the need to do away with the conventional power technologies in the context of global warming, and the scope for a large number of distributed energy sources spread all over the country, the investment focus in the integrated grid could shift from transmission segment to the distribution segment, though the importance of any of the segments may not diminish greatly. In the context of the social, environmental and economic issues being faced by our communities despite/ because of huge investments in the power sector, it is necessary to objectively consider the kind of power infrastructure for the future. It is safe to assume that the Civil Society will exert ever increasing pressure on the electricity companies of the future to become highly efficient, responsible, and law abiding. These companies are also expected to be aspiring to be amongst the best in the world. They are expected to be free from undue political interference and are managed through adequate professionalism in their ranks.

Keeping in view all the issues discussed in the earlier Chapters and sections, it will not be out of place to emphasise here the fact that there is no option for the country as a whole, and the state of TN in particular, to resolutely move away from conventional technology power sources, including coal, gas, diesel, nuclear and dam-based hydro as early as feasible, and that all sections of the society have to put concerted efforts in this paradigm shift. Such an energy transition to RE based power system is techno-economically viable, and is in the best interest of our communities.

If, as envisioned in this study report, the society goes for a large number of small size roof top SPVs OR wind turbines OR community based bio-energy/CSP type solar power plants the need for a stronger/ reliable integrated grid will increase, but the nature of the grid will be different. There can be very few conventional technology power plants such as few gas-based plants and dam based hydel plants, and pumped storage plants, which are already constructed and which have long life cycles. It is difficult to visualize many large-size power plants in operation by Year 2050, whether they are Solar based OR Wind based OR Bio-energy based. Large capacity (say in the size range of hundreds of MW) is most likely to be in very small number, if at all they are present. Most of the electricity in the system is likely to come from a large number of small size roof top SPVs OR wind turbines OR community based bio-energy/CSP type solar power plants: and such sources are preferably located very close to the existing power distribution network to avoid the construction of additional transmission lines. In the case of roof top SPVs this is entirely possible as most of the buildings can be expected to be located close to the existing distribution lines. In the case of off-shore wind parks, or a large solar power parks (such as in Rajasthan desert or remote areas of Gujarath), or if large size geo-thermal sites become techno-economically viable, some additional EHV lines may become necessary, but otherwise the emphasis is likely to be on strong and reliable distribution systems at voltages 33 kV and below. In order to supplement the power output from a large number of small size roof top/community based renewable energy systems, most of which may not be able to generate power when there is no sun light or wind, suitably designed CSP type solar power plants with heat storage facility for night time power generation can be expected to be installed at convenient locations such as each district/talukas.

It seems reasonable to expect that most of the loads in the future, except heavy loads like industrial, railway traction, construction sites etc., are likely to be fed by decentralized generation sources. Among the conventional technology power plants only hydel power plants of various sizes are likely to have presence: firstly because of the designed long life of the existing plants, and secondly because they may be seen as an integral part of the sustainable energy scenario. But due to increasing opposition to such plants on the grounds that there are social and environmental issues, it is unlikely that many additional hydro power plants of large size will be constructed between now and 2050. Gas based units may continue to have some relevance for few more years to bridge the gap till REs become fully mature. Most of the power produced in the large number of small size roof top SPVs OR wind turbines OR community based bio-energy/CSP type solar power plants will be expected to be consumed locally at the levels of distribution voltages such as 33 kV, 22 kV, 11 kV and below. Majority of electricity consumers are likely to be producers also (called as PROSUMERS) through roof top SPVs.

Micro Grids and Smart Grids

Another feature of the power grid of the future will be that it becomes an intelligent grid, where the monitoring, information flow and control functions are likely to be handled mostly by electrical/ electronic/pneumatic devices, than by humans. Such grids are already termed as 'Smart Grids', and the demand/supply of electricity at individual consumer level can be expected to be monitored and controlled much more accurately from remote locations (such as area/state load dispatch centres) through the automated usage of advanced communication and control mechanisms for optimal utilisation of the existing infrastructure as compared to the antiquated option of simply adding to generation capacity or load shedding. One obvious advantage of such a Smart Grid will be to the end consumers, because a Smart Grid must be highly reliable in its availability and in various parameters such as voltage, frequency and harmonics. A micro grid would consist of electricity consumers in a small geographical area (such as one or more villages; a residential colony in an urban area; or a colony of small-scale industries etc), many small size RE sources, and the deployment of smart and reliable metering, advanced protection, control and communication systems to manage the load and generation satisfactorily without having to depend on external sources for most of the time. Such micro grids can be connected to other similar micro grids, if considered necessary/essential, through advanced protection and communication tools. Reduced need for additional EHV/UHV lines and HVDC lines can also be visualized as very likely, because of the huge problems being faced in getting the right of way for such lines and substations. A recent example of how a 400 KV transmission line proposed between Karnataka and Kerala was delayed because of the sustained opposition by the local people in the hilly district of Coorg in Karnataka can indicate the seriousness of the problems associated with HV/EHV/UHV transmission lines. The power evacuation from the future power plants such as the proposed Cheyyur UMPP mayalso face similar problems.

<u>Bihar village declares independence from darkness and anonymity</u>

A good example of how a microgrid can address the energy accessibility problem in our villages is the Greenpeace's first solar-powered micro-grid in Bihar. The 100 kilowatt (kW) micro-grid currently provides an acceptable quality of electricity to more than 2,400 people living in Dharnai village in Bihar's Jehanabad district, near Gaya. Costing Rs. 3 crore, the solar-powered micro-grid is a comprehensive, first of-its-kind enterprise that provides 24x7 electricity to more than 450 households and 50 commercial establishments. This includes 70 kW for electricity generation and 30 kW for 10 solar-powered water pumping systems of three horsepower each. Built within three months, the quick-to-install micro-grid also takes care of 60 street lights, energy requirements of two schools, one health centre, one Kisan Training Centre (Farmer Training Centre) and 50 commercial establishments.

In the Indian scenario such microgrids will be highly suitable for providing electricity access quickly to about 33% of the population who have no such access even after 67 years of independence. These villages cannot wait any longer for the grid expansion to provide access to electricity. The micro grids are suitable to remote villages since a modest infrastructure can be set up rapidly, as quickly as in one week depending on the number of consumers. They can help in eliminating energy discrimination between rural and urban areas, as is happening with the integrated grid now. Such microgrids can enable villages and remote hamlets that are off the main grid to leapfrog into sustainable power access via solar PV (photovoltaic) mini-grids as a long-term solution rather thanas a stop-gap 'till the time the integrated grid reaches them.

The microgrid sector in India is dominated by smaller enterprises like Desi Power, Husk Power Systems, Saran Renewable Energies, Mera Gaon Micro Grid Power, Naturetech Infra, which all operate micro grids on a commercial basis. For example; Husk Power company provides light bulbs and a small amount of electricity to about 200,000 people in 300 tiny farming villages across the state of Bihar that have never been touched by the electric grid. Each village has a generator powered by burning and gasifying rice husks, a byproduct of farming that is otherwise wasted. Although no comprehensive statistics exist on the number of micro grids in India, a conservative count may indicate that they serve about 125,000 households in India, divided mostly between large, government-sponsored projects in the states of Chhattisgarh and West Bengal and private ventures centered on Uttar Pradesh and Bihar. The experience of such micro grids in India seems to be satisfactory, as evidenced by the estimation that about 125,000 households are being served, and that there are reports that even those villages which are connected to   the larger grid, but have highly unsatisfactory power supply, are seeking the services of such micro grids. Once the infrastructure in such micro grids become adequately strong, and if there are clear benefits to connect the same to the state network, such connection can be achieved through appropriate protection and control devices.

The state of TN should consider such micro grids to eliminate the energy poverty in rural areas. Since a minute control of electricity generated and consumed in such micro grids is essential, the next stage in development will be to make them 'Smart Grids' by enabling accurate information flow on critical parameters of the grid from all points, and to provide the facility for minute control of individual loads to match that of the total generation. This requires the usage of advanced communication, control and protection devices. It is envisaged that the future will have a very large number of such micro grids / smart grids with a provision to be connected to each other through the super grids by carefully implemented communication, control and protection systems. India Smart Grid Forum (ISGF) is a public private partnership initiative of Ministry of Power (MoP), Government of India for accelerated development of smart grid technologies in the Indian power sector. India Smart Grid Forum in consultation with India Smart Grid Task Force has formulated a comprehensive smart grid vision and road map for India, which is aligned to the Government's overarching objectives of "Access, Availability and Affordability of Power for All". Smart Grid Vision for India is to transform the Indian power sector into a secure, adaptive, sustainable and digitally enabled ecosystem that provides reliable and quality energy for all with active participation of stakeholders.

Conclusions:

The future of electricity scenario in the country will be determined by due diligence in effectively considering few major factors:

- The threats associated with the fast-looming Climate Change;

- The need to protect and enhance critical elements of nature such as air, water, soil, forests, rivers etc.;

- The need to reach a net-zero carbon emission scenario by 2070 or earlier, which necessitates to move away from fossil fuels as early as possible;

- To protect our communities from those kinds of electricity infrastructure needs which will force them to be displaced; and/ or subject them to health issues of pollution /contamination of air, water and soil;

- The economic sense to optimally harness the humongous potential in renewable energy sources within the borders of our own country;

- The need to minimize the overall societal level costs of the entire electricity infrastructure.

--

Chapter 3: Preparing for rising sea levels and its implications

India has an extensive coastline with a length of 6100 Kms on the mainland. The length of the coast line of India including the coastline of Andaman and Nicobar Islands and the Lakshwadeep Islands is 7517 Kms. Some of our most populous cities -- Mumbai, Chennai and Kolkotta, are situated on the coast. According to a report in the Telegraph in 2009, Ghoramara village located about 150 Kms South of Kolkata had lost almost 50% of the village lands due to the rising levels of the Hoogly River over three decades. This is also the fate of 7 lakh people from Malda and Murshidabad areas of Bengal who had to flee their villages submerged due to rise in sea level.

Large areas of India's coast line will be submerged due to sea level rise. According to a 1996 report by The Energy and Resources Institute (TERI), more than 5700 square kilometers along the coastal region will be inundated and almost 7.1 million of the coastal population will be directly affected. Several coastal areas will be submerged and the Lakshadweep Archipelago will be wiped off the world map. Thousands of acres of land in the Kutch, Mumbai and South Kerala regions will be submerged. Deltas of the Ganges in West Bengal, Cauvery in Tamil Nadu and many other rivers will be lost. The coastlines in Goa, Puri and Vizag have already receded to ridiculous levels due to sea-level rise, erosion, mass tourism and other human interventions.

According to the United Nations Environment Program (UNEP), India is one of 27 countries that will face the worst consequences of sea level rise caused by thermal expansion of ocean water due to global warming. According to estimates made in a study by Dr Gordon McGranahan, et. al in 2007, on behalf of the International Institute for Environment and Development, India has the second largest population in the world, of 63 million people located in the Low Elevation Coastal Zone (LECZ)2 with an area of about 82,000 sq. km. This area is extremely vulnerable to inundation due to sea level rise. There is sufficient proof that this must be an area of heightened responsibility for the Government of India, especially the Ministry of Environment, Forests and Climate Change.

Sagar Mala Project

As stated earlier, the ambitious Sagar Mala project may prove to be unviable in the face of the effects of climate change in the decades to come. This aspect of climate change has been well brought out by a team of experts from the National Maritime Foundation in the paper, 'Introducing Climate Resilience as the Fifth Pillar of the Sagarmala program'. The National Maritime Foundation has observed that the big-picture objectives of the SAGARMALA program for port-led economic development are highly commendable and very promising. However, the success of this ambitious program, the economies of Indias coastal states, and, in fact, the very fate of India's Blue Economy, rely heavily upon the long-term stability and security of the port-infrastructure over multiple decades. In view of this, there is an urgent need for government authorities at the centre, state, and district-levels to pay focused attention to ensuring resilience of port-infrastructure against the growing impacts of disasters such as floods, cyclones, and storm surges, compounded by sea-level rise and hotter ocean temperatures, all of which are being exacerbated by man-made (anthropogenic) climate change. Mumbai is a case in point. It is projected that by 2050, several parts of the city will be submerged, including such landmarks as the headquarters of the city's municipal corporation, the Reserve Bank of India, the Chhatrapati Shivaji Maharaj Terminus, Oval Maidan and Brabourne Stadium. The development of Mumbai port as a key component of the Sagarmala project cannot proceed if such projections are not evaluated and taken into account.

The total investments in developing the mega ports, the coastal economic zones and connectivity between ports and to the hinterland have been pegged at a whopping Rs 12 lakh crores. The Sagar Mala project is perhaps sound as a concept. However, given the reality of climate change, it may be prudent to proceed with caution. It is therefore disconcerting that the government plans to halve the previously estimated 10-year timeframe and fast track the work so as to complete over eighty percent of the project by 2025. The National Maritime Foundation has recently initiated a long term research to study the impacts of climate change on coastal and supporting infrastructure including supply chains. The study aims to develop a policy framework to incorporate climate resilience during planning stages of coastal development projects. It is hoped that accordingly the Government will incorporate climate resilience into the ongoing Sagarmala project even if it entails some years delay in the completion of the work.

Tourism infrastructure for Lakshwadeep, Andaman and Nicobar Islands

With regard to the massive tourism infrastructure planned for Lakshadweep and the Andaman and Nicobar islands, the very rationale defies logic. While the very future existence of Lakshadweep islands is in doubt, a report by the Intergovernmental Panel on Climate Change states that islands such as the Andaman and Nicobar might not be habitable in a few years due to rise in sea level and increase in climatic events like cyclones. Rather than pushing tourism and other development projects in such regions, the government could well plan and allocate funding for relocation and rehabilitation of these island populations that face dislocation due to Climate Change and rising sea levels

What is the Action needed?

Niti Aayog must make a thorough analysis of the impending threat of rising sea levels and prepare to relocate the huge mass of population as they move inland from coastal areas. It would be pertinent to note that the Government of Fiji has already taken steps to relocate coastal communities that are most vulnerable to rising sea levels. India may also have to cater to climate refugees from neighboring countries, due to sea level rise. Estimates point to 40 million South Asian climate refugees by 2050. India must initiate dialogue with other nations in order to come to an agreement on the quantum of climate refugees that could be accommodated by respective countries. This will eliminate much suffering of climate refugees fleeing to other countries in the years to come. It would also be important in the context of our internal security. Niti Aayog will also need to initiate a review of the Sagar Mala Project and the proposed tourism infrastructure for the Andamans and the Lakshwadweep Islands, against the back drop of rising sea levels

Chapter4. Zoning of the Indian landscape to prepare for climate change

Food, water and economic security with zone specific planning

India is a land beautifully crafted by the hand of God. The Himalayan Ranges and the Western Ghats are two marvels of the Indian landscape. The Himalayan ranges protect us from the cold Northern winds while the snow-fed rivers that originate in the Himalayas provide succor to the massive population inhabiting the Indo-Gangetic Plains. Towards the South and west of India, the forest-covered Western Ghats are vital to the South West Monsoons and provide much of the inflow to the life-giving rivers across the Deccan Plateau and South India, sustaining about a fourth of India's population.

Without the Himalayas and the Western Ghats, much of India would have possibly been a cold, wind-swept and arid landscape. It is more than obvious that these two regions are to be regarded as the principal Eco-regions of India and long-term policy measures are formulated and put in place to protect and safeguard the ecology of the Himalayas and the Western Ghats. The protection of these regions is vital in order to ensure food and water security for the people of India. Even industrial activities that the government prioritizes in order to propel the economy need huge quantities of water.

Can Make-In-India work if the environment fails?

The 'Make in India 'strategy cannot be sustained without adequate water for industries. For example, the manufacture of a single mid-size car requires approximately 1,50,000 litres of water! Therefore, environmental protection is not at the cost of development but is for ensuring long-term sustainable development and economic stability. Food and water security in their true sense also ensure health security by addressing the issues of malnutrition and water-borne diseases caused by river pollution at source and beyond. **Degradation of catchment areas and reduced water availability for downstream urban centers will result in over-dependence on bore-wells. This has already led to groundwater in Indian cities plummeting to levels where the water is dangerously contaminated with underground minerals and salts.**

The common line is that "Development cannot be stalled for the sake of Environment". All too often we come across those mouthing clichés like 'Development comes at a cost'. Our policy-makers have to introspect; what is 'Development 'and what is 'Cost'.

Can the Uttarakhand tragedy be written off as cost against Development? The 'development 'in the Kashmir Valley led to the encroachment of 85% of the Dal Lake and this was a contributory factor to the devastating floods in 2014.

Can the country thrive if catchment areas like Kodagu District in Karnataka are destroyed?

Western Ghats, Kodagu District, Karnataka

An example of degradation of a catchment area due to unplanned development,

Urbanization and loss of livelihood

Kodagu district in Karnataka is a case in point. Kodagu is a small district of 4,108 square kilometers, astride the Western Ghats. It is the principal catchment for River Cauvery and provides almost 50 percent of the total inflow. River Cauvery sustains about 80 million people across South India and also provides water for hundreds of major industries. It is obvious that protection of the Kodagu landscape and other river catchment areas is in the national interest.

Hence, it is a matter of deep concern that development in Kodagu is taking the form of unplanned rapid urbanization, and invasive tourism. Paddy fields and coffee plantations are fast being converted to residential layouts, sites, villas and tourist resorts. Meanwhile, development projects such as state highways, railways, power lines and hydro-electric projects are poised to rip through Kodagu.

A case of failed development

In 2018, torrential rains hit the hill slopes rendered vulnerable due to unplanned development. The result was unprecedented devastation in the form of floods and landslides. Many small farmers lost their homes and their lands were actually carried away in rivers of soil and water. Similar extreme weather events caused by climate change hit the hill areas of Kodagu again during 2019 and 2020.

Forest degradation and development infrastructure in forest habitats have resulted in heightened levels of Human Animal conflict in Kodagu, especially with regard to elephants and tigers. The situation is very serious with frequent incidents of deaths, injuries and crop loss. This will gradually force more agriculturists and workers to consider moving out and the land will be taken over for urbanization and town expansion.

Urban sprawl is gradually eroding food-producing areas in India. In Kodagu, productive agricultural land is being fast exploited by Real Estate and Construction interests. While financially satisfying for the investors, this has serious consequences on the local people and their lives. Sand mining, to meet the needs of the construction industry, is one example.

Sand-mining degrades rivers across the country as huge quantities of sand are sucked out of river beds. The Government's attempt to curb sand mining has had an undesirable paradoxical effect. The paradox here is that while it is perfectly legal to produce millions of bags of cement and boost the cement industry, the government now regulates the sand that will be required to mix the cement with! The Government has therefore created a Demand and Supply situation that has given rise to the 'Sand Mafia'. Sand is still available but at a price.

Measures to curb rampant urbanization and protection of food-growing regions from real estate interests will enhance food security and will also protect our rivers. It will also be crucial to identify alternative technologies for construction.

Policy-makers look at massive river-interlinking projects across the country in order to overcome drought. Instead, they must also act on less ambitious but more immediate river protection measures. The economic costs of the mega projects are mind-boggling while targeted regulation and prevention costs much less. In the same context, drought should be tackled through well planned watershed development schemes. This would be development in its true sense and spirit

Mountain ranges, deserts, forests and river systems have evolved through the millennia. The Kodagu example makes it clear that drastic man-made changes to natural geographic features will only cause irreparable harm to humanity. The country can well be drought-proofed by proper watershed management programs. The simple brilliance of Anna Hazare and Father Bacher in the Ahmednagar region of Maharashtra needs to be carried to the rest of the country. Existing watershed development agencies have to be revamped.

There needs to be in-depth analysis on the reasons why large parts of India remain prone to drought despite several watershed schemes. Watershed schemes must be transparent and monitored at the Gram Panchayat level. Awareness, capacity building and participation by Gram Sabhas need to be ensured.

Drought conditions in India are due to reduced inflow into the peninsular rivers originating in the Western Ghats, while flooding is caused by excessive water flow into the Eastern Himalayan Rivers. Both droughts and floods can be controlled by addressing the underlying causes in the water catchment areas, and by adopting benign measures such as water shed management, forest land restoration and reducing emissions of short life climate forcers (more on this later). **The costs will be marginal compared to the massive investment in river interlinking projects, many of which are likely to prove redundant when glacier melt will reach a stage when there is insufficient water to feed the drought prone rivers.**

The Prime Minister's plan to make the North Eastern States a global hub for organic produce is on target. This is precisely what needs to be done for the entire country. Each region or sub-region must be identified as a zone for a specific purpose with policies formulated accordingly. Principal Eco-regions [essentially catchment areas] and food-growing belts need to be kept relatively free from demographic pressures to the extent possible. The concept of zone-level planning if successfully implemented in India, can also be adopted by other countries. The proposed zoning could be under the following heads:

1. Catchment areas [Water producing Zones]

2. Food and Plantation areas

3. Population Centers

4. Industrial Hubs

5. Communication corridor

The previous Government at the Centre embarked on a program of identifying and encouraging Special Economic Zones with a slew of incentives for industries. Similar incentives would be required for communities living in Principal Catchment areas in order to provide them financial security so that they could make sustainable use of the landscape.

Some other countries already have the concept of Payment for Ecological Services [PES]. It is a simple idea of Upland Communities and Low land Beneficiaries. The College of Forestry at Ponnampet, Kodagu has prepared a document on the concept of PES for Kodagu [Reference Appendix A]. The system can be tried for Kodagu and then implemented across other catchment areas of the country. The Policies at the National Level should be formulated with a time-vision of around Year 2050. At that time-line, the policy makers can easily foresee the projected population with estimates of the productive and dependent percentage and the requirements by that time in terms of:

A. Water

B. Food

C. Energy

D. Housing

E. Education

F. Employment

G. Industries

H. Infrastructure in relation to the above factors.

Accordingly, the zoning process will need to be progressed.

Zoning must address the nature of the landscape and the resources it provides, with specific reference to river catchment areas

For the purpose of zoning, we cannot take a one cap fits all approach as India is blessed by very diverse ecology and natural resources across her landscape. Zoning can also play a role in dealing with displaced climate refugees both from within the country and from neighboring region. As mentioned earlier, the predicted scenario of Climate Change by the year 2050 can be broadly assessed and must be factor into planning.

Vulnerable populations from the coast will have to be resettled in other areas. Land must be identified for crores of people who will be displaced and infrastructure to include power and water has to be planned. It would be prudent to divert money from projects that will anyway be rendered redundant due to climate change and divert them to fund re-settlement. It should be ensured however, that these new population centers should not be set up in the fragile catchment areas and important ecological regions such as the Himalayas and the Western Ghats. In addition, keeping in view the massive loss of forest cover in India over the past few decades, forest land should also not be diverted for re-settlement.

CATCHMENT AREA POLICY FOR INDIA

Preamble

1. Water scarcity and drought are worsening across India and urgent policy measures are required to address the situation. It is for consideration that the Government will also have to cater to an increased in population -- from 1.3 billion currently to 1.6 billion by 2050. As our population rises, we will also be facing the severity of drought conditions worsening on account of climate change.

2. While considerable effort has been made for programs such as Ganga Rejuvenation and Mission Pani, there is an urgent need to identify and protect catchment areas across the country. The Principal catchments that drain the Indo-Gangetic Plains are across the Himalayan ranges while the Western Ghats, and the Vindhyas form the principal catchment areas of the rivers that sustain Peninsular and Central India. It would be very important for the Ministry of water resources to formulate a River Catchment area Policy for India in the interests of Water Security for the Country.

River Catchment Area Policy

3. The River Catchment area Policy for India will require the following parameters:

   A.   Identification of Principal Catchment Areas (PCAs)

   B.   Legal status of the PCAs — should be on par with that of protected forests.

   C.   Activities banned, restricted and permitted in the PCAs should be listed.

   D.   Incentives should be provided for the communities inhabiting the PCAs.

The concept of **Payment for Ecological Services (PES)** should be instituted for all PCAs across the Country.

4. The Western Ghats Ecology Expert Panel [WGEEP]

The WGEEP headed by Prof Madhav Gadgil was set up by the Government of India in 2010 and the Panel submitted its report in 2011. In essence, the report was a catchment area policy for the Western Ghats. Unfortunately, the report came under strong criticism due to political reasons and remained unimplemented. Subsequently, the High-Level Working Group on Western Ghats [HLWG] headed by Dr K. Kasturirangan was formed and the Working Group submitted its report in June 2013. If the recommendations of the WGEEP had been accepted with some modifications, it would have gone a long way in saving the river catchments in the Western Ghats. However, the recommendations of the HLWG are weak, and will not provide sufficient protection to the river catchments in the Western Ghats [Reference Appendix B].

5. In the light of the above observations and in order to provide water security to a larger population of people while taking climate change into account, it will be essential for the Ministry of Water Resources to convene a committee of experts at the earliest to prepare a draft Catchment Area Policy for India. In this context, the recommendations of the WGEEP could form the basis of interventions for the Western Ghats, while a similar panel would need to be formed to provide recommendations with respect to the Himalayas.

In addition to catering for displaced populations within the Country, contingency plans must also be made with regard to the possible massive influx climate refugees that may surge into India from neighboring Countries. Here we have to consider that the population of India would have peaked to about 1.6 billion by 2050. Hence the Government has to start preparing for:

   [a] The Natural increase in population by 2050,

   [b] The population of displaced people within the country due to sea level rise

   [c] The likely population of Climate refugees that may enter India.

The most urgent action to be taken by the Government of India and the State Governments is to ensure that there is restriction on migration to mega cities and especially to our coastal cities that are threatened by sea level rise. This will require the strengthening of village economies on a sustainable basis.

During Corona pandemic in 2020, millions of poor Indians migrated from cities back to their villages under the most tragic conditions. This is a clear indication that since Independence, we have failed to build a vibrant rural economy. Mahatma Gandhi said that the Future of India lies in its villages. The migrants' crisis of 2020 reminds us of his words. Decades after Gandhi, Anna Hazare championed the cause at Ralegan Siddhi .With water shed management as the main area of focus he strengthened the village community, fought corruption, brought prosperity to the villagers and reversed migration to cities. Flourishing villages will prevent rural populations from abandoning their agrarian base and shifting to far away cities to live in unclean slums

Forests are often linked with catchment areas. India has de-notified 700,000 hectares of forests during the past decade for development projects. Today, vast areas of our country reel under the specter of drought. Meanwhile, across the country, man-animal conflict is on the rise -- elephants, tigers, leopards, bears and gaurs are increasingly coming into direct contact with humans outside forest areas. The monkey menace is also a serious problem in many areas. There is growing anger and frustration among communities adjacent to forest areas and a constant demand on Forest Departments to capture, cull, trans-locate and barricade.

Unfortunately the focus is not on the root cause, i.e., habitat destruction. Media and celebrities line up for a 'Save the Tiger Campaign 'but as a nation we need to go further for improved forest management. [Reference Appendix C]

Chapter5. Stabilization of climate change in the Himalayas and the Mekong Basin

Climate change in the Himalayas is not India's problem alone. The Himalayas are home to over 100 million people including several indigenous communities. Increased temperatures will have a direct and drastic impact on over these communities whose livelihoods and culture are closely linked to the Himalayan Mountain Eco-system that touches many countries. As food and water security is threatened these populations will become increasingly dependent on food imports while also becoming more vulnerable to natural disasters such as avalanches and Glacial Lake Outburst floods.

Degradation of the Himalayan environment is a matter of deep concern to the entire global community and to the people of South Asia and South East Asia in particular. The Gangothri glacier that is the source of the Ganges has receded by 600 meters in the past 40 years. There has been a marked increase in the rate since 1971 and the glacier has been shrinking by 30 meters per year. According to ISROs Space Application Centre, as many as 127 glaciers of less than 1sq Km size have lost 38 per cent of their geographical area since 1962. The larger glaciers, which are progressively getting fragmented, have receded by as much as 12 per cent. Lester Brown of the World Watch Institute, USA says that due to the effects of global warming the pattern of precipitation in the Himalayas and the regions contiguous to the Himalayas will undergo a more drastic change in the years to come. This would give rise to a vicious cycle of more devastating floods and intense drought. He warns that due to reduced snowfall and glacier melt, ultimately there will be no water flowing in many of the Himalayan rivers during summer months whereas heavier rains will see more severe flooding during the monsoons, especially in the East Himalayan rivers that flow into India, China and the Mekong region. We are already witness to a cycle of more pronounced drought and floods in the region.

The Himalayas stretch across a length of 2500 Kms and averages a width of 300 Kms spanning across India, Nepal, Bhutan and China and Pakistan. India's future prosperity in this area is tied to an environment that is shared with our neighbors.

This region is vitally important not just for these countries but many others in South Asia and South East Asia, as it is the source of a number of rivers sustaining one-fifth of humanity.

The Mekong Basin is home to almost 65 million people and grows the rice to feed almost 300 million people in the world. The Mekong along with other important rivers draining South East Asia and China also originate in the Himalayas. The Himalayas and the Mekong Basin are interdependent ecological regions. The severe degradation of the Himalayan region from climate change is therefore a matter of deep concern to the entire global community and South and South East Asia in particular. The Mekong Basin, contiguous to the Eastern Himalayas, faces the maximum impact of the degradation of the Eastern Himalayan glaciers that are melting at a faster pace compared to the Central and Western Himalayas.

While the Himalayan waters sustain the Mekong basin, the pollutants - Black Carbon, (BC), from this heavily populated region have an adverse effect especially on the Eastern Himalayas. The effect of BC in the Eastern Himalayas is more profound, since the areas of the Mekong Basin have some of the highest population densities in the world, resulting in greater industrial and automobile emissions that generate more Black Carbon. Scientific studies have revealed that the rapid glacier melt and reduced snow fall in the Himalayas can be addressed through drastic reduction of regional green house gasses and emissions of Short Lived Climate Pollutants especially Black Carbon (soot). Science has also established that forests and vegetation play a major role not only as sinks for CO2, but also in absorbing massive amounts of Short Lived Climate Pollutants including Nitrous Oxides, Sulphur Dioxide and particulate matter like Black Carbon. Therefore, reduction of Black Carbon in the Mekong basin together with Forest Land Restoration to improve the health of the ecosystems will mitigate climate change both in the Himalayan region and the Mekong basin. Tree planting and development of Parks in certain cities and industrial hubs should also be undertaken in the Himalayan region and the Mekong Basin.

In his book 'Fire and Ice 'Jonathan Mingle describes the tragedy of Kumik, the remotest village in Ladakh -, in the Zanskar region, which was sustained by the water melt of a glacier in the vicinity of the village. However, with the gradual disappearance of the glacier due to climate change and ill-timed and unpredictable snow melt, the villagers were finally forced to relocate and to abandon the village that had been their home for centuries.

It is evident that if the trend of reduced snowfall, increased precipitation and shrinking of the Himalayan glaciers continues, the result will be catastrophic for several millions of people in South Asia, South East Asia and China. Food productivity of the entire region would be severely affected due to the cycle of droughts and floods. It has been estimated that during this century, the accelerated ice melts in the Himalayas flowing into the seas will cause sea levels to rise by I meter. Such rise in sea levels would destroy fifty percent of the rice fields of Bangladesh. It would also result in millions of 'climate refugees 'fleeing from the low-lying areas in India, China, Bangladesh, Indonesia, and Vietnam.

The rise in temperatures will adversely affect the biodiversity of the Himalayas. As regards the riverine ecology, the degradation of biodiversity will be felt not only in the Himalayan region, but along the entire course of the rivers and up to the estuaries where they drain out into the oceans. Changes in the river regime will impact aquatic biodiversity and will threaten the livelihoods of innumerable communities that depend on fishing.

Sustainable development in the Himalayan region should be guided by reducing the threat to the Himalayan Eco system while ensuring financial security of the local communities:

- o Increased self-resilience of people though ecosystem management. Greater community involvement in forest management through Sustainable Forest Management practices [SFM]

- o Increased self-resilience through alternate energy sources – biogas, micro hydro energy and eco friendly cook stoves
- o Providing financial security through schemes for Payment for Ecological Services [PES]

Development projects can look for cooperation with various other organizations such as the Asia Pacific Forestry Network, Climate and Clean Air Coalition and the Global Alliance for Clean Cook Stoves and the Centre for People and Forests [RECOFTC].

Regional emissions and pressures in the Himalayan region

Like any other Eco-system, the Himalayas are adversely influenced by emission of green house gases in any part of the world. However, it is crucial to understand that there is considerable influence from the emissions of certain green house gases and aerosols right from within the Himalayan ranges and contiguous areas. These local emissions create a regional climate impact that combines with overall global warming to further accentuate the temperature rise.

Being a snow covered mountain Eco-region the Himalayas are particularly vulnerable due to the trapping effect of the valleys. Moreover, Black Carbon Aerosols with a shorter life span remain in the atmosphere for longer periods in cold climates. It is relevant to note that the rapid melting of ice caps in the Arctic is influenced not only by Global Warming but also due to regional emissions -- mainly from Eurasia, as well as oil and off-shore oil exploitation. Heavy shipping traffic with large concentrations of Nitrogen Oxide emissions is another cause for the Arctic Haze that compounds the effect of Global Warming. The Siachen Glacier is another case in point. According to a study by the WWF, the past two decades has seen a rapid melting of the glacier and it is amongst the fastest melting glaciers in the world. During 1984, the Siachen glacier dispute flared up between India and Pakistan. India's genuine security concerns in the region have resulted in   massive troop deployment by both countries. The study states that the Siachen Glacier has been melting alarmingly more due to military activity of India and Pakistan than due to global warming.

Sources and causes of regional emissions in the Himalayas

1. Asian Brown Haze

The Asian Brown Haze is caused mainly by domestic wood and coal fires, vehicle exhaust fumes obsolete brick kilns, etc. Certain mega-city hotspots such as Delhi, Beijing, and Dhaka have been identified that contribute significantly to the Black Carbon in the Asian Brown Haze. The Asian Brown Haze is causing a regional heating effect that is accelerating the glacier melt in the Himalayas. It is estimated that the heating effect of the brown haze is the same as that of the Global Warming due to Green House Gases. In a sense, the Himalayan region is perhaps being subjected to a 'Double Whammy 'due to the combined effect of overall Global warming coupled with the impact of the Brown Haze. Black Carbon is an important component of the Haze and reduction in Black Carbon emissions should be given top priority. According to an IGSD/INECE report, the impact of Black Carbon on melting snow-pack and glaciers in the Himalayas may be equal to that of CO2.While the effect of Carbon Dioxide emissions has a more global effect, there is sufficient scientific evidence to prove that other non-CO2 gases and aerosols effect the climate in the immediate vicinity of the emissions.

2. Urban Heat Islands

Both the core Himalayan region and the contiguous areas have a number of large cities and townships that form Urban Heat Islands [UHI]. The UHI effect is like a balloon of higher temperature formed over the urban areas. This balloon of higher temperature is shifted to the adjoining non-urban areas due to wind factors and causes a higher temperature in these UHI contiguous areas outside the cities/towns. The effect is more intense in mountainous regions due to the 'trapping 'effect of valleys -- as can be seen in the high levels of pollution in the Kathmandu Valley. Levels of air pollution in Kathmandu are one of the highest in Asia, although the number of vehicles is far less than Mumbai and Delhi. UHI effect can extend to a range of up to 2.4 times the size of the city, beyond the city limits. Hence increased urbanization in the Himalayan region can create a number of climatic 'hotspots 'that contribute to the overall rise in regional temperature. Planned tree planting with selected quick growing tree species and shrubs can diminish the effect of Urban Heat Islands.

3. Pressures of Tourists and Pilgrims

Tourists and pilgrims form a large floating population in the Himalayas, concentrated in the cities and popular tourist/religious destinations. They exert a more direct 'point 'influence and contribute to the Urban Heat Island effect in cities such as Srinagar and Kathmandu. This effect is also pronounced in site-specific pilgrimage destinations such as the Gangothri Glacier that is the source of the Ganges River.

4. Military presence in the Himalayas

Troop deployment and movement of administrative army convoys result in heavy emissions in the core areas of the Himalayas.

It is emphasized that this paper does not attempt to make a case for demilitarization of the Himalayan region, since troop deployment is linked to security concerns and national policies of the concerned countries. However it is pointed out that there is ample scope for the respective countries to take practical steps to see that there is considerable reduction in the emissions caused by army deployment and administrative convoys.

Cost Benefit Analysis

It is beyond the scope of this paper to present a detailed cost-benefit analysis of reducing emissions [principally black carbon and non-CO2 Green House Gases] in the Himalayas. However, available information point towards huge savings by way of improved health conditions, [especially of women and children], and savings in the energy sector. Other benefits include the avoidance of disasters caused by climate change, such as bursting of glacial lakes due to increased levels of glacial melt. Huge benefits would also accrue by preventing climate-induced drought/floods in the lower regions such as the Indo-Gangetic plains, Southern China and Bangladesh. Regional strategies for mitigation of Climate Change in the Himalayas and adjoining regions will address key issues such as food productivity and water security for large parts of South Asia, South East Asia and China. This would again lead to reduced tensions within the region. A review of the economics of climate change by the Government of the UK states that, if no action is taken now, the overall cost and risks of climate change could be equivalent to the loss of 5% of global gross domestic product (GDP) each year. If a wider range of risks and impacts is considered, the estimated damage could reach as high as 20% of global GDP.

The Time factor and the Regional perspective

Leading climatologists have warned of the need to act immediately to cut greenhouse gas emissions, with a window of 10-15 years for global emissions to peak and decline, and a goal of at least a 50 percent reduction by 2050. However, it is to be understood that from the regional perspective of the Himalayan Ecology, we do not have even so much time. The Himalayan region will witness increased population pressures in the coming decade. This is all the more reason that emission reduction strategies must be executed at the earliest. Concentrated efforts must be made to drastically reduce aerosol and non-CO2 emissions within the next five years, primarily to cut down on the formation of the Brown Haze over the Himalayas. If this is not done, the ecological damage to the Himalayas and especially the Himalayan Glaciers may be irreversible.

There is no time to be lost in carrying out further exhaustive research and analytical studies. Findings of credible research studies already carried out need to be taken into account. The stress should be on identifying and categorizing the principal sources of aerosol, Black Carbon and non-CO2 emissions such as Nitrous Oxide. This will need to be followed by working out strategies for achieving required scale of reduction for these emissions within a mutually agreed time-frame. Forest land restoration and judicious tree planting in selected cities and industrial hubs will have to complement the reduction in emissions.

THE TIME FACTOR

It is of utmost importance to emphasize that we do not have the luxury of time. The project will need to be implemented well before 2030 or else we may have crossed the Tipping Point. If reductions in Short Lived Climate Pollutants (SLCP) are delayed until 2030, it will be more difficult if not impossible to keep warming under 2°C by the end of the century.

Permafrost melt is already underway in the Arctic region and it has also been observed in the Tibet region. This will result in release of massive amounts of CO2 and Methane into the atmosphere. This has not been factored into the projected rise in global temperatures. Therefore the need for urgent action is all the more crucial.

It is also important to note that Black Carbon deposition on snow and ice will accelerate permafrost melt and set off a vicious cycle of temperature rise With regard to Permafrost melt, it is an extremely precarious situation when it comes to melting of sub-sea permafrost due to warming of the sea water. This will accelerate the release of methane and Carbon Dioxide into the atmosphere

As mentioned earlier, there is no time to be lost in carrying out further exhaustive research and analytical studies. The stress should be on identifying and categorizing the principal sources of aerosol, Black Carbon and non-CO2 emissions such as Nitrous Oxide.

[Please see Note on Permafrost Melt and its Concerns at Appendix D]

THE HIMEK ALLIANCE PROPOSAL

STABILIZATION OF CLIMATE CHANGE IN THE HIMALAYAS AND THE MEKONG BASIN

A case for India's leadership and regional cooperation in the Himalayas and the Mekong Basin

It has been established that the Himalayan glaciers are melting rapidly, especially in the Eastern Himalayas. Therefore the proposed HIMEK Alliance must be formed at the earliest and the various programs for reduction in Black Carbon emissions and Forest Land restoration programs should be implemented with a sense of utmost urgency. Procrastination, lack of political will and bureaucratic delays will push us beyond the tipping point and will result in immense human suffering.

The HIMEK Alliance proposal was first drafted by Col CP Muthanna during 2009. He took up the subject with International Union for Conservation of Nature (IUCN) India and IUCN Asian Regional Office (IUCN ARO). This was followed up by a number of consultations and discussions at various forums in India, China, Thailand, Nepal and Bhutan. Two key developments were the discussions at IUCN ARO during March 2012 and a two Hour presentation at the IUCN Regional Conference at Bangkok during September 2015.Other important events included a presentation during the IUCN World Conservation Congress at Hawaii, during 2016 and another presentation during the IUCN Regional Conference at Islamabad during 2019.

While the Himalayan waters sustain the Mekong basin, the pollutants from this heavily populated region would have an adverse effect on especially on the Eastern Himalayas. Therefore, reduction of Black Carbon (BC) in the Mekong basin together with Forest Land Restoration to improve the health of the ecosystems would have a positive effect in mitigating climate change both in the Himalayas and the Mekong basin.

For India, the Himalayas and the Western Ghats are threatened with irreversible damage to their ecology and await their sad fate as we are led by our immediate self-interests in depleting our natural resources and degrading our environment. This is our Tragedy of Commons, but it is not too late if we can act now with a sense of urgency. The Himalayas in particular has immense opportunity for international cooperation for shared Eco-regions. India can well lead the way.

During the consultations on the HIMEK Alliance proposal, there was active participation of delegates of the concerned countries and other agencies including IUCN, FAO, UNDP, Chinese Academy of Sciences, TERI, WHO, AIT, RMFN Asia, RECOFTC, GB Pant Institute, etc. There was consensus on the importance of reducing BC emissions and for large scale Forest Restoration programs in the Himalayas and Mekong Basin through Regional cooperation. Col Muthanna has also had consultations with the Indian Army with regard to reduction in Black Carbon in the Himalayas.

In a very important development, during 2018, IUCN [Asian Regional Office] and the Asian Institute of Technology, both based in Thailand have signed an MOU on the basis of the HIMEK proposal [Reference Appendix G]. A HIMEK Cell has also been set up at the IUCN India office, New Delhi during December 2021

THE WAY AHEAD

1. The Eleven countries of the Himalayas and the Mekong Region, together with certain other global agencies should form an organization [HIMEK Alliance] to formulate and execute a joint strategy for mitigation of climate change in the Himalayas and the Mekong basin. Such an organization could be modeled along the lines of the existing Arctic Union.

2. Advantages of Regional Cooperation on the Himalayas

A joint strategy by the eleven countries will have tremendous advantages. It will ensure that there is an integrated, time-bound approach to tackling the issue with the active involvement of other concerned International Agencies. In the case of global climate change negotiations there was reluctance by developing countries to come to agreement with the terms and conditions set by the Western countries. However, in a regional alliance of this nature it will be easy to obtain the cooperation of all the countries as it will be a Win-Win situation.

Parameters for a joint strategy frame work

• Identifying the extent of the zone requiring intervention. This would include the Himalayan ranges and contiguous areas including the Mekong basin. Broadly, the Himalayan Ranges would be the core zone and the contiguous areas would be the outer zone. Initially, the outer zone could be for a radius of fifty Kms from the core zone and the Mekong basin. The outer zone could then be increased periodically till a maximum laid-down radius is covered under the action-plan.

• An analysis of the interventions required in the core zone and the outer zone in order to reduce emissions and mitigate climate change and to stabilize the effects of global warming to the extent possible. The countries concerned will then have to sign an agreement on the various interventions and the time frame within which these will be implemented.

• The process will need to be facilitated by other organizations such as UNEP, RRCAP, GREEN CLIMATE FUND, GEF, GLOBAL ALLIANCE FOR CLEAN COOK STOVES, etc These organizations should also be involved in organizing the required funding mechanisms.

Recommended interventions for the HIMEK Alliance - Some with higher sensitivity and higher priority in the core zone compared to the outer zone.

a. **Industries:** Certain industries must be banned and phased out. Alternatively, they should be permitted only on introduction of upgraded technology that will sufficiently minimize emissions. They will also require financial assistance to incorporate cleaner technologies. There is good scope for reducing BC emissions by improved technology for thousands of brick kilns in the region. Nepal, Bhutan and Bangladesh may need financial support for installing cleaner technologies. International funding will be required.

b. **Automobiles**: Automobiles in both the core zone and the outer zone should convert to environment friendly fuel. As far as the Government of India is concerned, priority for converting to CNG or LPG should to given to Jammu, Dehradun, Srinagar, Shimla and Manali. These cities should also get priority in setting up facilities for charging of electric vehicles.

All the countries concerned maintain a very large military presence in the core zone. Thousands of army trucks move within the core zone every day-to cities in the Himalayan region such as Jammu, and Manali, etc., Eco-friendly fuel for these vehicles is essential. However, in view of logistical considerations, this may not be practical in the near time. Hence the Himalayan countries must ensure that military vehicles plying in the Himalayan region confirm to required emission norms. Adequate mass transport facilities such as buses should be provided for tourists and pilgrims.

A simple cost effective innovation developed and patented by Somender Singh, a Mysore-based technician is also available for reducing vehicular emissions and improving fuel efficiency.

c. Road Construction activity

There is constant road construction and maintenance activity in the Himalayan region, using obsolete construction methods that require the burning of large quantities of coal tar. This contributes substantially to the Green House gas and Black Carbon aerosol emissions in the Himalayas. There is an urgent need to introduce cleaner technologies for road construction and repair in the Himalayas.

d. **Demography**: Demographic pressure always translates to greater levels of human activity. Concentrations of populations should be avoided. The Governments should encourage well-planned satellite townships in the periphery of the Himalayan region, rather than increased growth of cities such as Kathmandu, Jammu and Shimla.

e. **Forest land restoration**: An intensive forest land restoration program by all the countries is vital. The establishment of trans-boundary National Parks would improve the management of forests along border areas. Ecological Territorial Army Battalions comprising of ex-servicemen will be able to play a very important role in forest land restoration in the Himalayas in India. The concept of Ecological Territorial Army units could be shared with other countries of the HIMEK Alliance

f. **Improved technologies in domestic fuel consumption**: There is excellent scope for improved technologies for domestic fuel consumption in the Himalayan region. The National Program on

Improved Cook stoves [NPIC] in Himachal Pradesh is an example. This delivers the benefit of emission reduction combined with improved health of women and children.

g. Land use practices: Burning of huge agriculture residue each year such as paddy in the North Indian State of Punjab and in some Himalayan states can have a direct influence on the on the increase in Black Carbon emissions and the Brown Cloud over the Himalayas. The fires from these fields often spread to the Himalayan foothills causing forest fires. The Himalayan landscape is also stressed due to over grazing and slash and burn cultivation in Eastern Himalayas. Well thought out strategies will be required to mitigate these pressures on the Himalayas.

h. Mega-City hot spots: There are a number of mega cities in close proximity to the Himalayas. These include Delhi, Kolkota, Dehradun, Dhaka, Karachi, Quetta, Xinning , etc. There will be a need to reduce pollution levels, especially of Black Carbon from these cities. Moreover, further migration to these cities and their expansion should be discouraged and other population centers should be developed further away from the Himalayas.

HIMEK ALLIANCE COUNTRIES

As a result of the consultations, discussions and presentations, there is a broad agreement in forming a regional cooperation, at present termed as the HIMEK Alliance. The following eleven nations would be involved:

Himalayas: India, China, Pakistan, Nepal, Bhutan, Bangladesh

Mekong Basin: Myanmar, Vietnam, Laos, Cambodia, Thailand

Apart from IUCN, there would be a need to involve various other agencies including UNEP, UNDP, RECOFTC (Regional Community Forestry Training Center for Asia and the Pacific) and RRCAP, ie., Regional Resource Centre for the Asia and the Pacific [Asian Institute of Technology]. RECOFTC and RRCAP [AIT] would be especially relevant as partners in the project. RECOFTC has wide experience in carrying out Community Based Forest Land Restoration in the Asian Region, while RRCAP is focused on reduction of Black Carbon and other Short life Climate Pollutants.

Ongoing efforts and important initial steps that have been taken are as follows:

(a) IUCN Document.

In 2011, IUCN ARO published a document titled "Stabilization of Climate Change in the Himalayas: Strategy for a Regional Response". The document was based largely on the concept note by Col CP Muthanna [Reference Appendix E]

(b) Trial project on Cookstoves at Balkila in Indian Himalayas

The Environment and Health Foundation has donated 40 fuel efficient cook stoves to IUCN India about two years ago to be distributed to economically weaker families in the Balkila Watershed of Uttarakhand in the Indian Himalayas. The feedback from the families is very positive.

(c) Innovation by Garuda R&D for petrol and diesel vehicles

The Garuda R&D is based at Mysore in South India. The founder of the organization is a technician named Somender Singh. He has developed and patented an innovation for petrol and diesel vehicles. Mr Somender Singh has been in touch with Indian Government authorities. This technology results in drastic reduction of Black Carbon emissions and other Green House gases. It also improves fuel efficiency considerably. [Reference Appendix F]

(d) Project involvement of the RECOFTC

RECOFTC has carried out extensive work in Community based Forest Land Restoration in South East Asia. Representatives from RECOFTC have participated in a number of discussions on the HIMEK Alliance proposal and they have expressed their willingness to work as an implementing partner in the HIMEK Alliance project.

(e) IUCN India-Forest ROAM Restoration program

IUCN India in partnership with the World Research Institute [WRI] and the G.B. Pant National Institute of Himalayan Environment & Sustainable Development (GBPNIHESD), piloted the Restoration Opportunities Assessment Methodology (ROAM) project study in Uttarakhand State in the Himalayas. The key findings of the study were released in 2018. It is planned to replicate this model across seven countries - Peru, El Salvador, Mexico, India, Kenya, Viet Nam and Uganda. This impactful model can be adapted towards Forest Land Restoration in the HIMEK Alliance project.

f. **MOU between IUCN Asia Regional Office and Asian Institute of Technology**

During 2018, The IUCN Asia Regional Office and the Asian Institute of Technology have signed an MOU on the basis of the HIMEK Alliance. Delegates from Asian Institute of Technology [AIT] have attended discussions on the HIMEK Alliance proposal. As a result of the discussions, it was decided that the two organizations would cooperate towards the implementation of the HIMEK Alliance proposal. It was mutually agreed that while IUCN would facilitate Forest Land Restoration programs, the Regional Resource Centre for Asia and the Pacific [RRCAP] which is part of AIT would facilitate reduction of Short life Climate Pollutants [SLCPs], with special reference to Black Carbon. Accordingly an MOU was signed between IUCN ARO and AIT in 2018, with the aim of cooperation in taking forward the HIMEK Alliance proposal. [Reference Appendix G]

BROAD WORK PLAN FOR PHASE I [FIRST FIVE YEARS]

The program will have to be in the nature of a coordinated project between Himalayan Region countries - India, China, Pakistan, Nepal, Bhutan and Bangladesh and Mekong Basin countries - Myanmar, Vietnam, Laos, Cambodia and Thailand.

The phase I of the implementation will need to commence after adequate consultations have been held by IUCN ARO and funding agencies have been confirmed.

Pilot projects would normally be necessary. However to save time these could be replaced by evaluation and assessment of the existing projects on BC emissions and Forest Land restoration in the region. For example, projects in South and South East Asian countries are already under way with regard to Forest Land Restoration, fuel efficient cook stoves, and upgrading brick kiln technology. This evaluation and assessment process could be done along with the consultation and planning process

Hence the overall project will be scheduled as follows:

Preliminary Phase: Consultations, cost estimates and planning; duration of one year

Execution Phase

Stage 1: Duration of 3 years

Stage 2: Duration of 5years. The various phases may not be strictly time bound and may spill over in time and in geography.

Forest Land Restoration- 1000 hectares per country

As regards the direct implications for the forests in the Himalayan region, mitigation strategies are vital in order to ensure that the forests themselves do not get degraded and destroyed due to climate change. In the middle altitudes of the Himalayas, Chir Pine is taking over Oak dominated forests. Degradation of natural forests due to invasive species and other climate associated factors will accelerate climate change and the rise in temperatures will in turn result in further degradation of the forest Eco-systems. In this context it is very important to take urgent measures to check the trend of rising temperatures in the Himalayas. There has also been severe loss of forest cover and degradation of natural forests in both the Himalayan region as well as the Mekong basin. Hence it will be vital to launch a massive forest land restoration program to regain natural forest cover in the Himalayas and the Mekong Basin.

The forestry related part of the project will be taken up through Community Based Forest Land Restoration/Landscape restoration [CBFLR/CBLR]. The program could be under the overall coordination of RECOFTC. For added benefit of addressing water security concerns, the attempt will be to carry out the CBFLR/CBLR projects in important river catchment areas. Prevention of forest fires should be inherent to the Forest land Restoration programs.

If there is paucity of resources, Forest Land Restoration could be 1000 hectares for the Himalayan countries and 500 hectares for the others. Forest Land Restoration in Bhutan may not be required due to the excellent forest cover in that country.

Note: In the case of India, China, Pakistan and Bangladesh, it is envisaged that the Forest Restoration and emission reduction programs will be in areas that are in the Himalayan region or in close proximity to the Himalayas.

Emission Reduction

The main sources of black carbon emissions in the Himalayas and the Mekong basin could be categorized as follows:

- House-hold emissions due to inefficient cook stoves
- Obsolete technology for brick kilns
- Vehicular emissions
- Obsolete road building methods in the Himalayas
- Forest fires
- Burning of agriculture crop residue
- Emissions from military sources

Brick Kilns: All obsolete brick kilns to be phased out. In India, Green Tech Solutions has been carrying out excellent work on improved technologies for brick manufacture. Existing legislations and projects on brick kilns in some countries could be adopted by other nations in the Alliance [Reference Appendix H].

Vehicle emissions: In the case of mega cities close to the Himalayas and large cities in the Mekong region could provide incentives for electricity and CNG operated taxies. There could also be a law by the end of the first five year period to make it mandatory for all new taxis to run on electricity or CNG.

Each mega city should have a team of trained technicians for vehicle modification through the Garuda R&D technology. This will be subject to evaluation and assessment of the technology by the concerned countries. The CCAC may be approached to act as the agency for the initial evaluation. The bus transportation companies could be considered for the initial modifications.

Road construction and maintenance: Obsolete methods for road construction and maintenance result in high levels of black carbon emissions in the core areas of the Himalayas. This is due to preparation of coal tar. There is a need to replace these obsolete technologies and eradicate this source of BC emissions in the Himalayas

Forest Fires: The Canadian Forest Department has excellent expertise on early warning systems and prevention of forest fires, and their expertise could be procured on payment of a fee, to train forest department personnel on fire management for every country including Indonesia. Indonesia is being considered due to the massive forest and peat fires that have an influence in the Atmospheric Brown Clouds affecting the Himalayas and Mekong Region. The other very important agency with regard to prevention of Forest Fires is the FAO. The FAO Regional Office for Asia Pacific could play a very important role. China has developed an efficient fire detection robot with a radius of 15 Kilometers. One such system has been deployed for forest fire detection in Cambodia. It would be very important for China to share this technology with all the other HIMEK Alliance countries

Burning of Residual agriculture crop: This issue will need to be addressed by creating awareness among farming communities in the region. Alternate methods of agriculture and soil fertility management will need to be promoted to the farmers.

Conclusion

This is a broad way forward in order to initiate the planning process. It is emphasized that the work plan will need to be finalized after detailed analysis and consultations.

1. There is sufficient evidence to indicate that regional emissions of Non-CO2 GHGs, aerosols and Black Carbon are a key factor in the rise of temperatures in the Himalayan region. Reducing these emissions will result in mitigating the overall effect of climate change in the Himalayas. A massive forest land restoration program across the Himalayan Region will also be essential to curb the current trend of reduced snowfall and rapid glacier retreat.

2. This could best be achieved by a system of regional cooperation of the eleven Himalayan and Mekong basin Countries, along the lines of the existing Arctic Union. This regional cooperation is termed HIMEK Alliance. International agencies will need to facilitate the process.

3. The concept for cutting down on regional emissions, combined with efforts to improve the forest cover could herald a new dimension to dealing with Climate Change. There is scope to use this strategy in the Arctic region, the Alpine region and the Andes. Reducing BC emissions from the heavily industrialized regions surrounding the Arctic can help to slow down snow and ice melt in the Arctic and this will also help in retarding the rate of permafrost melt. In the case of sub-sea permafrost melt and warming seas, there is a strong case for reduction in BC emissions from the industrial hubs located in the coastal regions across the Globe.

4. International agreements such as the Kyoto Protocol, the Bali Declaration and the Paris agreement call for achieving desired level of reduction in emissions by 2050. The Himalayas and other similar Eco-regions do not have the luxury of time until 2050 and need urgent action to formulate and execute a joint strategy for the Himalayas and the Mekong Basin before it is too late.

APPENDICES

Possibilities of Payment for Ecosystem Services (PES) in Kodagu, Western Ghats of India
(Abridged version)

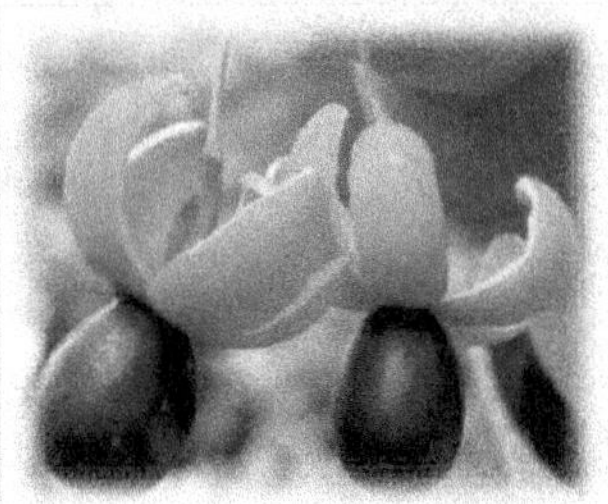

College of Forestry, Ponnampet – 571216 , Kodagu, University of Agricultural Sciences, Bangalore, India

Canadian Model Forest Network, Ontario, Canada

Project Team

Principal Investigator: Dr. N.A. Prakash, Dean (Forestry), College of Forestry, Ponnampet. University of Agricultural Sciences (UAS), Bangalore

Biodiversity:

Dr. Mohana G.S. Junior Rice Breeder, Agricultural Research Station, Ponnampet (UAS, Bangalore); **Dr. Sathish, B.N.** Assistant Professor, Department of Forest Products and Utilization, College of Forestry, Ponnampet (UAS, Bangalore); **Dr. Smitha Krishnan**, ETH, Zurich.

Carbon:

Dr. Devagiri, G.M. Associate Professor and Head, Department of Natural Resource Management, College of Forestry, Ponnampet (UAS, Bangalore); **Dr. Devakumar, A.S**. Associate Professor, Department of Forestry and Environmental Sciences, UAS, GKVK, Bangalore; **Dr. Philippe VAAST**, Senior Researcher, CIRAD and World Agroforestry Centre, Nairobi, Kenya

Hydrology:

Dr. C.G. Kushalappa, Professor and Head, Department of Forest Biology and Tree Improvement, College of Forestry, Ponnampet (UAS, Bangalore); **Mr. Raghu, H.B**. Assistant Professor, Department of Forest Biology and Tree Improvement, College of Forestry, Ponnampet (UAS, Bangalore); **Col. C.P. Muthanna**, Secretary, Kodagu Model Forest Trust, College of Forestry, Ponnampet.

Economic Evaluation:

Drs. M.G. Chandrakanth, T.N. Prakash, B.V. Chinnappa Reddy, K.B. Umesh and **P.S. Srikanthamurthy,**
Department of Agricultural Economics, UAS, Bangalore

2012

© Forestry College, Ponnampet, UAS, Bangalore,
IndiaEmail: kushalcg@gmail.com

Cover Page design: Dr. Mohana, G.S., ARS,
PonnampetPrinting: Type Corner, Bangalore

Contents

Preface

Kodagu is one of the greenest landscapes in India and is part of the Western Ghats, a global hotspot of biodiversity. With 81% of the geographical area under tree cover, district harbors diverse ecosystems such as natural forests, sacred groves, coffee agroforestry systems and paddy fields that contribute to the diversity of species which represent 8% of India's plant wealth. However, the landscape and demography in Kodagu is currently undergoing rapid changes which are bound to impact the ecosystem and services that flow from the district and in turn communities within and outside.

World over including many developing countries, Incentive Based Mechanisms like Payment for Ecosystem Services (PES), Ecological Certification and Landscape Labeling are being promoted as means of "Green and Clean development". Under the current scenario in Kodagu, there is an urgent need to formulate an action plan for economic development model based on sustainable utilization of natural resources by adopting above approaches. We have essentially attempted here to review the key ecosystem services from the landscape of Kodagu and look at mechanisms of providing incentives to communities. We wish to acknowledge Canadian Model Forest Network for providing financial assistance and Kodagu Model Forest Trust, Ponnampet for facilitating this initiative. Our special thanks to farmers and elected representatives of Kodagu district who have actively participated in consultation process and given their valuable inputs for this endeavor. We hope this document will raise awareness among key stakeholders involved in providing and receiving the life sustaining ecosystem services and pave ways for development on the principle of ecological economics.

Dr. N.A. Prakash

Dean (Forestry) and Team Leader, PES
ProjectCollege of Forestry, Ponnampet,
Kodagu University of Agricultural
Sciences, Bangalore

Introduction

In recent times, developments in applying economic thinking to the use of biodiversity and ecosystem services have increased manifold. There is also a compelling cost-benefit case for public investment in ecological infrastructure (especially restoring and conserving forests, river basins, wetlands, and others), particularly because of its significant potential as a means of adaptation to climate change (TEEB 2010). Another dimension is that payments for ecosystem services are generating considerable attention because they have the prospects to create new funding opportunities for biodiversity protection and ecosystem services that contribute to human wellbeing. In this connection, United Nations initiative in the form of Conference on Environment and Development, has developed a concept of accounting the natural resources and termed it as "Green accounting" under System of National Accounts (SNA). The recent Convention on Biological Diversity Conference of the Parties (COP-10 in Nagoya, October 2010) led the global players to declarations on making the use of environmental goods part of the national accounting and the next convention to be held 2012 in Hyderabad, will offer an excellent opportunity for India to showcase the national strategies and potential for PES in the country.

An overview of ecosystem services : Valuing Ecosystem services and to get payment from the beneficiaries are generally termed as PES schemes and they focus on ecological/environmental services provided by forest conservation, reforestation, sustainable utilization, agrobiodiversity and agroforestry, for which there is an existing market demand, or for which such demand can emerge in future under appropriate conditions.

Need for PES Mechanism in Kodagu: Natural landscapes of Kodagu district provides an excellent opportunity for promotion of the recently emerging concept of the 'green economy'. The diverse ecosystems and associated diversity of bioresources of Kodagu have contributed to economic development of the region. In addition to these direct benefits, there are many ecosystem services provided by the landscape such as provisional, regulating, cultural and supporting services. In recent years, the district has become an important tourist destination owing to its landscape beauty and places of cultural interest. There can be many such examples of contribution of ecosystems for the development of "Green Economy" in the district as indicated by one of the highest developmental indices among the districts in India. Higher green cover and associated higher economic and developmental index is a proof of the synergies between sustainable environment management and sustained economic development. The ecosystem services provided by the "Greenscape" of Kogadu is not only confined to the district. A range of life supporting and sustaining benefits are provided to communities downstream the river Cauvery in terms of water, climate regulation and, timber and fuelwood needs. Thus it is imperative to promote sustainable management of the landscape that contributes to protection of environment through incentive based mechanisms like PES.

The Kodagu Landscape: The district is one of the highly wooded regions in India with 81% of the landscape under tree cover representing all the major tropical forest types. The tree covered landscapes include 1214 sacred forests under community management, shade grown coffee plantations under private ownership covering 33% of the landscape of the district and producing 38% of India's coffee. With diverse ecosystems and associated species, Kodagu has been identified as a micro hotspot of biodiversity within the larger Western Ghats region. In addition to hosting spectacular biodiversity, the forested ecosystems provide a range of ecosystem services which sustains the livelihood of the local communities.

Landscape dynamics: The forested landscape is currently undergoing dynamic changes with respect to land use and land cover. This diverse multistoried agroforestry system is undergoing transformation with respect to canopy density and diversity due to changes in the production systems under the current liberalized market situation. Coffee plantations are becoming more open and native trees are being replaced by exotic Silver oak. Natural forests under the management of forest department are also subjected to biotic pressures and considerable proportion of the natural forests have been converted to Teak and other plantations. Sacred forests are being encroached and degraded. Paddy lands are either left fallow or being converted to other land uses or habitations. Unregulated tourism is resulting in impacts on the environment and natural resources.

Against this background, an attempt to compile and quantify the key ecosystem services namely biodiversity, carbon sequestration and watershed services based on the scientific studies carried out in the district was undertaken. This document is an abridged version of the policy document which is envisaged to be an eye opener to general public and private entrepreneurs to support the cause of Incentive Based Mechanisms for sustainable landscape management which are already in practice elsewhere in India and world.

Literature cited:

TEEB, 2010. The Economics of Ecosystems and Biodiversity: Mainstreaming the Economics of Nature: A Synthesis of the Approach, Conclusions and Recommendations of TEEB (Bonn: United Nations Environment Programme).

Background of the study

This being the first effort from Kodagu to collate information on ecosystem services and to develop policy document on PES through stakeholder participation, following events were held:

Brainstorming session held on 22nd November, 2011 identified Biodiversity, Carbon Sequestration and Hydrological services as the key services from landscapes of Kodagu and these services need to be evaluated for PES. Based on these discussions, a proposal was developed and submitted to Canadian

Model Forest Network (CMFN) by the University of Agricultural Sciences, Bangalore and a Memorandum of Understanding was signed between the partners on 25th January, 2012.

A team of researchers from Agro-Paris Tech, France collaborated with the project team in field data collections on stakeholders perception on PES which was shared in the interactions held on 23rd February at Ponnampet and 24th February,2012 at Bangalore. The event in Bangalore was attended by Vice-chancellor of the University and other senior officers.

Project team members explained the proposal to stakeholders in the first meeting held on 31st January, 2012 at College of Forestry, Ponnampet. Second stakeholders meeting held on 13th February, 2012 to review the progress of work and to include valuable inputs provided from the stakeholders in the policy document.

The draft policy document was presented to invited service providers and receivers of ecosystem services from Kodagu through a buyers-sellers meet held on 13th March, 2012. About fifty delegates representing different stakeholders like farmers' self help groups, women self help groups, Agricultural Scientist Forum of Kodagu, Kodagu Model Forest Trust, Kodagu Planters Association from 'Service Providers Group' and representatives from tourism sector and industries representing the 'Service Users Group' attended the meeting. Fruitful interactions were held and valuable inputs were collected on the perception and mechanisms of providing PES.

The final interaction with representatives of Kodagu Zilla Panchayat and three Taluk Panchayats was held on 14th March, 2012. The presentations made by the project team was helpful in creating awareness among the elected representatives on ecosystem services and how we could use them as a tool in economic development of the district. In his address, president of Kodagu Zilla Panchayat strongly advocated the need for providing PES from landscapes of Kodagu and assured all the support for taking the policy document to the regional and national government.

Biodiversity vis- *a- vis ecosystem* services: A perspective for Kodagu district

Sathish, B. N., Mohana, G.S. and *Smitha Krishnan*

Kodagu district is one of the most important landscapes within the Western Ghats biodiversity hotspot region harboring all the major tropical forest types of India. It has more than 81% of its land under tree cover (Forest Survey of India, 2011) and hosts 8% of India's and 35% of Karnataka's floristic diversity, which incidentally includes the largest tree of the country. Further, it is a major contributor to the coffee production at the national level and there are 34 distinct land tenure and tree management systems unique to this region. The district has diverse ecosystems such as Natural forests including grasslands and sholas, Sacred groves, Coffee agroforestry systems and, Paddy and wetlands. These ecosystems not only supports rich biodiversity but also play a very important role in providing timber, non timber forest products, fuel wood, fodder etc. and indirect services in terms of major sinks for carbon, soil and water conservation, nutrient recycling and regulating hydrological cycle.

Biodiversity in natural forested ecosystems: Natural forested ecosystems cover an area of 46 per cent of the total geographical area of the district. These ecosystems include evergreen, semi evergreen, moist deciduous, dry deciduous and scrub forest types. Among these, evergreen forests form the major vegetation type covering 33 per cent of the total forested landscape followed by moist and dry deciduous forests (5 per cent each). Evergreen forests also include small proportion of high altitudinal grassland sholas mainly occurring on western aspect on either part of Ghats crest. Structure of the forests in terms of density and basal area indicates that the mean number of stems in Kodagu were comparable with the forests in the Western Ghats.

Tree species richness range between 100 to 174 in low elevation evergreen forests, 90 to 126 in medium elevation evergreen forests and 90 to 100 species in high elevation evergreen forests. In moist deciduous forests,less than 50 and in dry deciduous forests around 30 trees species are reported. These richness values are comparable with other forests in the Western Ghats. In addition to rich diversity, evergreen forests are also considered as treasure house of endemic species. For instance, 48 per cent of the species are endemic to low elevation evergreen forests of Kadamakal reserve forest in Kodagu (Pascal and Pelissier, 1996). Among these endemics, Dipterocarpaceae members are dominant in number with *Dipterocarpus indicus* and *Vateria indica* representing 21 per cent of the trees (Elourd *et al.,* 1997; Pelissier, 1997). Shola forests or high altitudinal evergreen forests forms a unique ecosystem. These forests are characterized by small patches of forests in the valleys surrounded by grasslands. These forests support unique species diversity and hosts variety of orchid species. Sixty one species of orchids belonging to 32 genera were recorded from Tadiandamol, the highest peak in Kodagu, of which 46 species are epiphytic and 16 are terrestrial (Rao, 1998).

Biodiversity in cultural landscapes – Sacred groves: These are the unique traditional landscapes which play a vital role in biodiversity conservation. Kodagu has 1214 sacred groves covering an area of 2550 hectares. The density of sacred groves is very high i.e. one grove for every 300 ha of land, which is highest density in the world. Every village in Kodagu has at least one and in many cases more than one sacred grove. There are 39 villages, which have more than seven groves each. Considering the wide diversity of vegetation features, deities worshipped and communities involved in protection, Kodagu can be regarded as 'hotspot' of sacred grove tradition in the world. (Kushalappa and Bhagwat, 2001). Several studies on the floristic structure and diversity conducted in sacred groves indicate that they are as good as natural forests. Species richness was higher in larger sacred grove compared to the smaller groves and regeneration status of plant species increased with the size of the grove (Tambat, 2001). Higher number of plant species with medicinal and other utility value from coffee plantations and sacred groves are reported when compared to the reserve forests (Bhagwat, 2002).

Biodiversity in Agro-ecosystems:

Coffee based agroforestry systems: These systems in Kodagu are the examples for human managed systems and represent remnants of original forest, hence potential areas for conserving diverse flora and fauna. Studies have indicated that the canopy cover, floristic composition and structure are mainly determined by the management of plantation and original forest type. In recent years, there has been a shift in choice of shade trees from native to fast growing exotics such as *Grevillea robusta* and hence there are coffee based agroforests with only native shade trees, a combination of native and exotics and few dominated by only exotics. Despite these changes, the agroforests have retained a high diversity and density of shade trees (Bhagwat, 2002; Sathish, 2005; Krishnan, 2011). They are acting as corridor between the fragmented natural forests and facilitating gene flow, seed dispersal and pollination services. In addition, they also play a crucial role in sequestering carbon and regulating hydrological cycle as evidenced by the outcome of the CAFNET project (2011).

Status of Biodiversity in the coffee agroforestry systems in Cauvery watershed area: Findings of the CAFNET project (2011) showed that structure of the shade cover of the coffee estates in Kodagu is quite complex. The tree density is high, compared to remnant forest patches (for example *sacred groves*), and highly variable across estates. The observed richness is about 280 different tree species and the estimated richness could be 320 species. Among the different trees *Grevillea robusta* (Silver oak) represents close to 20% of the trees of the Cauvery watershed but 57% of the trees belong to miscellaneous native species indicating that system still has multi species composition. An illustrative filed guide of trees in the coffee based agroforestry system has been published by the CAFNET project team gives the detail account of tree diversity (Poornika *et al.*, 2011). In total, 42 epiphyte species were recorded from the coffee agroforestry systems. The proportion of epiphytic species significantly increased up to certain thresholds of canopy cover (75%) after which it decreased. Coffee under evergreen ecosystem supports higher population of bacteria and Arabica coffee harbors more bacteria compared to Robusta. Fungal population was high in evergreen ecosystem and number of *Actinomycetes* was more in evergreen ecosystem, with Robusta harboring higher number. Lignin decomposing and nitrogen fixing bacteria were higher in coffee grown under evergreen ecosystem whereas cellulose decomposers, starch hydrolyzing and pectin utilizing bacteria were higher in deciduous ecosystem.

Paddy lands: Paddy is the major staple grown in the district in an area of about 35000 ha in valleys. Compared to other irrigated areas of the state, the cultivation regime is still traditional i.e. local varieties are grown under low input systems (Mohana, 2010). Rice fields not only serve the purpose of provisioning services such as food and fodder, but also provide a means of water infiltration owing to huge volumes of standing water for at least 3 to 4 months. Paddy fields also harbor a number of weeds in addition to faunal diversity elements such as crabs, frogs and other lesser known insects. However, a comprehensive study is lacking in floral and faunal diversity in the paddy cultivation systems.

Faunal Diversity: As discussed in the above sections, the landscape is very rich in flora which in turn supports higher faunal diversity. The district has one National park and three wildlife sanctuaries. The Rajiv Gandhi National park is one of the best managed national parks in the country with high density of Asian Elephants and tigers. These protected areas also hosting relatively high density of other faunal species. The district has one of the largest numbers of Asian elephants in the country. In the recent years due to reduction in habitat size and quality, increasing Human Elephant Conflicts (HEC) have been recorded. Narasimnan (2004) has described close to about 310 birds in the district. Under CAFNET project (2011) studies have indicated the presence 109 bird species, 6 small mammals and a diverse group of micoflora. Results have also indicated that the bird communities react negatively to high levels of *Grevilea robusta*, with loss of richness and diversity. As the proportion of Silver oak increases in an area, the total number of bird species decreases. However, this trend is observed only after the proportion of Silver oak exceeds 20-30% of the total trees which is the threshold to support the diversity as well as accrue revenue to the farmer. Further, the Coffee plantations also provide ideal habitat for small Indian Civet and Civet coffee which is very popular in many countries could also be specialty coffee from Kodagu. The relevance of coffee estates as corridors for large population of elephants is also indicated from the study.

The remnant forests (sacred groves as well as privately owned remnant forests) provide important nesting and forage resources to many species of bees and butterflies. Krishnan (2011) has listed about 82 morpho-species of bees of which four are social bee species and 78 morpho-species of solitary bees belonging to 12 genera from three families. Further 78 species of butterflies (excluding *Lycanidae* family) have also been reported from the district.

Table 1. Documented Status of Biodiversity in different Ecosystems of Kodagu

Biodiversity elements	Number of species	Source
Plant species	1342	Keshavmurthy and Yoganarasimhan (1990)
Tree species in Coffee agroforestry systems	280	CAFNET report (2011)
Birds	310	Narasimhan (2004)

Birds in Coffee Agroforestry systems	109	CAFNET report (2011)
Snakes	49	Sathish (2009)
Frogs	23	Dannial (1998)
Small mammals	6	CAFNET report (2011)
Orchids	67	Keshavmurthy and Yoganarasimhan (1990)

Biodiversity – The need for PES: However, in recent years there are major changes happening in the different ecosystems which could have an impact on landscape level biodiversity (*please see the landscape dynamics section in Introduction chapter*). In view of these, a landscape approach has been strongly recommended for pursuing conservation of biodiversity in Kodagu. For successful conservation, it is essential to include all the ecosystems and local communities in conservation planning and to take care of their welfare. In this context, Incentive Based Mechanisms such as Payments for Ecosystem Services (PES) become crucial.

Valuation of Biodiversity (with inputs from B.V. Chinnappa Reddy, UAS, GKVK, Bangalore)

Identification and valuation of ecosystem services (ESS) from different components of biodiversity is a prerequisite and the starting point for determining the payment for ecosystem services (PES). After prioritization of ecosystem services, it is possible to identify potential benefits with and without protection or conservation or management with different regimes.

Soil and water conservation values: After taking the costs of programs to conserve the resources of water and soil (this can be termed as avoided costs) as proxy for valuation of this ESS, the value of these resources works out to be at an average of Rs. 200 per acre per year. This will be paid by the society. The aggregate value of this service for a catchment or degraded landscape can be arrived at by aggregating these values for entire landscape scale. The mode of collection (payment vehicle) for this service could be in the form of tax or levy on agricultural commodities, at the rate of 2 percent on the value of commodity.

Another ESS that is closely linked to this is the reduction of sedimentation rate in reservoirs and availability of additional water to farmers for a longer time. The cost of desiltation of deposited silt can be considered as the proxy for value of this service from biodiversity. The additional benefit in terms of agricultural output that could be realized will be substantial and this value could be the basis for determining the PES. Again 2 per cent of this value can be taken basis for estimating ESS value.

Non wood product values of biodiversity: Form the experience gained in valuation studies conducted in Muthathi forest area, we feel that 1-2 percent of value of these services can be considered as payment towards ESS. The payment mechanism could be a tax (1-2%) on the commodities or products originating from the biodiversity.

Nutrient availability and recycling values: Presence of rich biodiversity enhances microbial and below ground biomass activity often resulting in increased availability of native nutrients or recycling of nutrients. A Study carried out in Coorg (Shruthi *et al*, 2010) on the process value of phosphate solubalising bacteria (PSB) estimates the value of magnitude of nutrient recycled as Rs. 237- 1167 per ha. This works out to a mean value of Rs. 700 per ha, as nutrient recycling value of below ground biodiversity. Farmers can save this much on fertilizers if they take efforts to maintain the below ground biodiversity. Further, this is an organic source of nutrients, therefore, the agriculture products obtained through this process command a premium price. Thus, the average PES for this service could be worked out as 2% of this value which is Rs. 14 per ha. The payment vehicle for this service could be the cess at the rate of 2 percent on the products originating from the Kodagu region.

Value of timber: Value of this service is quite high compared to other ESS, as a result of ready market for lumber/timber. A considerable effort is involved in the making available this service, hence, we need to take into account the net value of providing this service. At 2 per cent rate, the payment to the timber product service (provisioning service) works to about Rs. 530 per ha at lower end (lower value timber) and Rs. 8340 per ha (high value timber) at the upper end. The revenue collection mode (payment vehicle) that could be used here is cess on the timber products sold.

Recreation values of Biodiversity: Assuming about 10 percent of recreational value as the payment for this service, the PES for this service is about Rs. 27000 per ha. The payment vehicle in this case could be the entry fee to the recreation site.

Value of Pollination services: The incremental yield from pollinating service of bee can be quantified and a percent of it (2 %) can be considered as the PES. We can consider the net value of coffee output per acre as the basis for determining the PES value at 2 percent.

These values are only indicative ones, because the Kodagu district is rich in biodiversity. However, to arrive at exact values of ESS and concomitantly to make PES, studies need to be conducted exclusively for Kodagu district.

Payment Mechanisms for Biodiversity Conservation (Inputs from Dr. Mohana, G.S., Sathish, B.N. Smitha Krishnan and Anand, M.O.)

Agency driven biodiversity conservation: This involves buying/adapting a piece of land or area by either private or public agency exclusively for conservation purposes which will be done in association with the forest department. The land might have already an established biodiversity element or the procured land might be utilized for reinstating the native biodiversity elements of the past. The ownership of the land will not change but rights to access and management can be worked out after a thorough discussion with the forest department. This approach can be profitably utilized in Kodagu as there are many areas with unique biodiversity. For instance, there are areas where *Dysoxulum malabaricum* (an endangered species with very high timber value) is naturally available. This agency driven strategy can also encompass supporting initiatives aimed at diversity conservation by an individual or community. Kodagu owns a large number of privately owned forests or diverse coffee agroforests which harbor considerable biodiversity. These are retained by the farmers over generations, thus maintaining the original vegetation cover. Provisions for payment to the owner to sustain this interest are desirable but the transaction costs of such incentives should not be colossal. Such conservation easements could also be provided to the community in case of community protected forests, mainly the sacred groves.

Payment for Access to Species or Habitat: An area with research significance or potential for bio-prospecting can be made accessible to research agencies or pharmaceutical companies on an agreed payment regime under the purview of Forest department. Biodiversity Management Committees (BMC) under Grama Panchayaths can also receive payments for access to plant resources within their village limits. Even this option can be advantageously employed in Kodagu district as areas where medicinal and other plants of research are available in plenty. For instance, *Mappia foetida,* a plant that yields compound " Campothecin" used for curing cancer is available in forests of Kodagu. Another unique product from the landscape which can be explored under this mechanism is the anti-obesity chemical obtained from *Garcinia gummigutta.*

Eco-tourism: Kodagu offers overwhelming opportunities for eco-tourism with spectacular array of places having rich biodiversity and are of cultural significance. It has already made a mark in eco-tourism worldwide. Home stays that offer the glimpse of traditional food are becoming popular throughout the district. However, there is a pressing need to systematically nurture this venture further for higher economic benefits of the local community.

Eco-certification and Geographic Indications (GI) tag: Towards these, there are immense opportunities in Kodagu and in recent years, many certification schemes such as eco-certification and possibilities of bird friendly coffee, elephant coffee, civet coffee and forest certification could be explored. Recently two products (Citrus and Cardamom) have obtained GI. Another new incentive tool "Landscape labeling" could also be very relevant for Kodagu.

Literature cited:

Bhagwat, S. A. 2002. Biodiversity and conservation of cultural landscapes in the Western Ghats of India,

    Ph.D. Thesis, University of Oxford, United Kingdom.

CAFNET India Final report, 2011. (http://www.ifpindia.org/Managing-Biodiversity-in-Mountain- Landscapes.html)

Dannial., 1998. Frogs of Kodagu, A Research report p 82.

Elourd, C., Pascal, J.P., Pelissier, R., Ramesh, B.R., Houllier, F., Purand, M., Aravajy, S., Moravie, M.A. and Gimaract – Carpentier, C. 1997. monitoring structure and dynamics of a dense moist evergreen forest in Western Ghats.

Forest Survey of India, 2011. State of forest report. URL Forest Survey of India www.envfor.nic.in.

Keshavmurthy, K.R. and Yoganarasimhan, S.N. 1990. Flora of Coorg (Kodagu), Karnataka, India. Vimsat publishers, Bangalore.

Krishnan, S. 2011. Pollinator services and coffee production in a forested landscape mosaic. Department of Environmental Sciences, ETH Zurich. *Ph D thesis.*

Kushalappa, C.G. and Shonil Bhagwat, 2001. Sacred groves: Biodiversity, threats and conservation. In. Skaankar, U.R., K.N. Ganeshaiah and K.S. Bawa. (eds) *Forest genetic sources: Status, Threats and Conservation strategies.* Oxford and IB, New Delhi, p 21-29.

Mohana, G.S. 2010. Genetic diversity of rice in the central Western Ghats: Prospects of conservation and utilization, Published in *Book of abstracts: First Indian Biodiversity Congress IBC-2010*; National Seminar 28-30th December 2010, Thiruvananthapuram, Kerala. page 18

Narasimhan, S.V. 2004. Feathered Jewells of Coorg. Coorg Wildlife Society. pp. 1-171

Pascal, J.P. and Pelisser, R. 1996. Structure and floristic composition of a tropical evergreen forests in south west India. *Journal of Tropical Ecology,* 12 (2): 191 – 214.

Pelissier, R. 1997. Heterogente spatiale et dynamique d'une forest dense humide dans les Ghats Occidentaux de l'inde. Publications du Department d' *Ecologie 37*, Institute Francias de Pondichery.

Poornika Rani, B.J., Sathish, B.N., Mohan, G.S., Somanna Chitiappa and Kushalappa, C.G. 2011. Field guide – *Trees of coffee Agroforestry Systems of Kodagu,* CAFNET project, Forestry College, Ponnampet – 571216, Karnataka, India pp. 260

Rao, T.A., 1998. Conservation of Wild Orchids of Kodagu in the Western Ghats. Navabarath, Bangalore.

Sathish, B.N. 2005. Assessment of tree diversity in coffee plantation under different land tenure systemsin Virajpet taluk, Kodagu. *M.Sc Thesis, University of Agricultural Sciences, Bangalore.*

Sathish. 2009. Snakes of Kodagu. Coorg Wildlife Society. pp.1-143

Shruthi, H.R., Raghavendra K.M.. Usha, K.S., Myna, C.C., Padma, Chinnappa Reddy, B.V. and Balakrishna,

A.N. 2010 . Economic valuation of Ecosystem services by below ground biodiversity : an analysis of selected provisional services of P sexualizing microorganisms. In Ecological Economics – An approach towards Soci-Economic and Environmental sustainability. Eds. Sunil Nautiyal and Bibhu Prasad Nayak, K. ISEE pp. 75-81

Tambat, B. 2001. Vegetation composition and reproductive ecology of few tree species in fragmented landscapes (sacred groves) of Kodagu, Central western ghats, *M.Sc. Thesis, University of Agricultural sciences, Bangalore.*

Carbon Sequestration Potential and PES Feasibilities in Kodagu Landscape

Devagiri, G.M., Devakumar, A.S. and Philippe VAAST

The atmospheric concentrations of Greenhouse Gases (GHGs) such as carbon dioxide, methane, carbon monoxide, ozone etc. have increased in the post-industrial era, primarily due to the combustion of fossil fuels and human induced land use and land cover changes. This has resulted in increased amount of heat trapped in the atmosphere and consequently gradual warming of the earth's surface. The Intergovernmental Panel on Climate Change (IPCC) has projected a rise in temperature of 0.6 to 4^0C in the coming decades. Such global climate change will have enormous impacts ecosystem functions and services.

Vegetation stores carbon in the form of biomass through photosynthetic process by sequestering atmospheric CO_2. Indian forests serve as a major sink of CO_2. It is estimated that the annual CO_2 removal by India's forest and Trees Outside Forests (TOF's) is enough to neutralize 11.25 per cent of India's total GHG emissions. In response to the climate change, Kyoto Protocol came into existence during 1997 and the countries which are signatory to the protocol agreed to a 5.2 per cent reduction in emissions of GHGs by 2012. India ratified the protocol in August, 2002 and is committed to reduce the atmospheric concentration of GHGs. It is important that such agreement provides incentives not only for Reducing Emissions from Deforestation and Forest Degradation (REDD and REDD+), but also for Sustainable Management of Forests (MoEF, 2009).

In light of the above facts, the present document is aimed at quantifying the carbon sequestration potential of the Kodagu landscape and to devise payment mechanism that suits the landscape since the district has 81 % tree cover and thus play a crucial role as sink of CO_2.

Carbon sequestration in different land use types of Kodagu: The total carbon pool from the major components of carbon sinks of natural forest of Kodagu (Table 1) was found to vary from 207 t ha^{-1} in sacred groves to 77 t ha^{-1} in dry deciduous forest, while evergreen forest had slightly lesser carbon pool of 170 t ha^{-1} as compared to semi-evergreen forest and sacred groves (Devakumar, 2009). These estimates are comparable with the values from other tropical ecosystems.

Table 1. The total carbon from different components of forest and sacred groves in Kodagu district

Vegetation type	Above ground biomass(t ha^{-1})	Litter biomass(t ha^{-1})	Herb biomass(t ha^{-1})	Soil carbon(t ha^{-1})	Total carbon content (t ha^{-1})
Evergreen	115	2.11	3.05	50.07	170
Semi Evergreen	112	1.72	2.11	63.12	179
Moist Deciduous	48	0.13	0.13	42.00	90

Dry Deciduous	43	0.10	0.18	33.45	77
Shola grassland	61	1.92	4.73	56.13	124
Sacred groves	140	0.38	--	67	207

Results from studies undertaken in CAFNET project (http://www.ifpindia.org/Managing-Biodiversity-in-Mountain-Landscapes.html) showed that coffee agroforests composed of Arabica shaded by either native or exotic tree species sequestered carbon at the same rate as reference forest (Table 2). To a lesser extent, this also appears to be the case for Robusta shaded with native species. With values in the range of 138-206 t ha^{-1}, the total carbon sequestered in the present coffee systems are well above the median carbon sequestration potential of other agroforestry systems estimated at 95 t ha^{-1} in the tropical regions.

Table 2. Above ground biomass and carbon distribution in coffee agroforests of Kodagu as compared to adjacent natural forest Carbon (t ha^{-1})

Land-use system	Tree layer	Coffee	Soil	Litter	Total
Natural Forest	97	--	97	2.4	196
Arabica native	88	4.8	112	1.6	206
Arabica exotic	73	3.3	105	2.2	183
Robusta native	78	13.0	90	1.8	182
Robusta exotic	47	10.1	78	1.9	138

Above ground biomass and carbon pool at landscape level

A recent study conducted by Devagiri *et al.* (2011) has revealed that Kodagu district contributes 70 per cent of total above ground biomass and carbon pool quantified from three other adjoining districts in the Western Ghats region of Karnataka. Above ground biomass ranged from 0.05 to 250 t ha^{-1} with a mean density of 92 t ha^{-1}. While the vegetation carbon density in the coffee agroforestry systems ranged from 0.03 t ha^{-1} to 120 t ha^{-1} with a mean density of 44 t ha^{-1}.

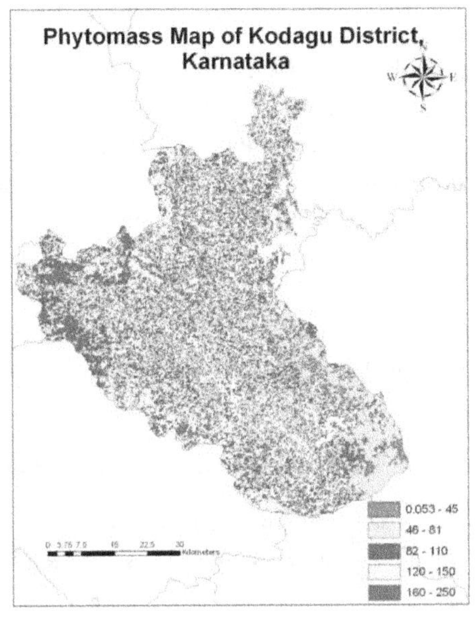

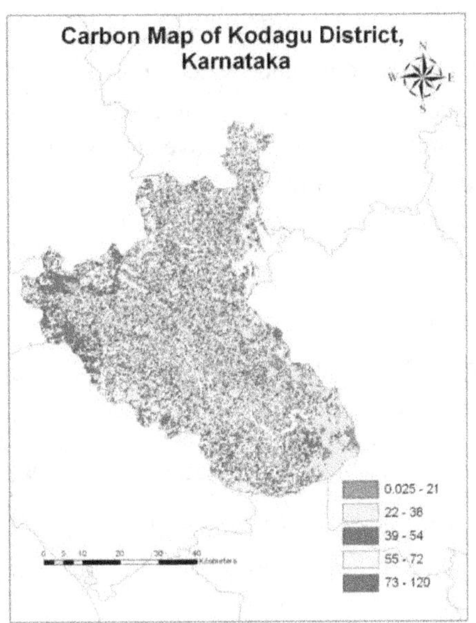

Valuation and payment mechanism: Economic valuation of carbon sequestration service offered by the natural forest ecosystem or any wooded ecosystem is essential for successful implementation of PES mechanism. There are many efforts around the world to value forest or tree based land use system as a source of carbon sink and for their contribution to mitigate global climate change. From the literature it was observed that the values range from 5 to 125 USD per tC. Studies conducted in Kodagu landscape have revealed the carbon sequestration potential of different land use types ranges from 40-150 tC ha^{-1}. Even if we assume modest estimate of 90 tC ha^{-1} sequestered by the vegetation, it amounts to 9000 USD or Rs. 40,500 per hectare of vegetation at an assumed rate of 10 USD per tC in the international market.

Possible models that could generate carbon credits for Kodagu landscape includes Best Management Practices (BMP), Afforestation and Reforestation (ARR) and avoided deforestation or forest degradation through REDD and REDD+. Under BMP, coffee agroforestry systems will qualify and the suitable mechanisms are CDM and REDD+. It is now widely recognized that sink related CDM projects can promote sustainable development and resilience of the smallholders' production systems. The Bio-carbon Fund (www. biocarbonfund.org) established by the World Bank, is a prominent source of funds for such projects. Another option is to utilize the benefits of Forest Certification (FSC), eco-certification and landscape labeling for sustainably managed landscapes like coffee agroforests. Activities that increase biomass accumulation in community forests such as village forest committees (VFC) areas and sacred groves and, in forest plantations and natural forests could be brought under REDD and REDD+ mechanisms as envisaged under Green India Mission of Ministry of Environment and Forests, Government of India

Proposed payment vehicles: As regards to taxes as payment vehicle for carbon, it is envisaged to collect Green Tax from new vehicles as one time tax equivalent to the amount of life time tax which can be collected at respective Regional Transport Offices and the distribution of the collected tax should be made based on the per cent of forest cover in respective districts. The payments for carbon must be based on carbon sequestered per year rather than carbon stocks and this could be linked high end car companies and other carbon emitting industries.

Conclusion: Studies undertaken in Kodagu district provide enough indications that the landscape as whole has high carbon stock and sequestration potential. Carbon sequestration service from the land use system consisting of tree cover is a well accepted service for Incentive Based Payment Mechanism to the land owners. Further, in the current context of global climate change, there is an urgent need to formulate PES mechanism for Sustainable Carbon Management (SCM) to halt further degradation of the landscape.

Literature cited:

CAFNET India Final Report, 2011. (http://www.ifpindia.org/Managing-Biodiversity-in-Mountain-Landscapes.html)

Devagiri, G. M., Money, S., Sarnam Singh, Dadhawal, V. K., Prasanth Patil, Anilkumar Khaple, Devakumar, A.S. and Santosh Hubballi. 2011. Assessment of above ground biomass and carbon pool in different vegetation types of south western part of Karnataka, India using spectral modeling. *Tropical Ecology* (in press).

Devakumar, A.S. Devagiri, G.M. and C.G. Kushalappa. 2009. Study of carbon dynamics in natural forests of Kodagu. Project report submitted to DST, GOI, (project no. ES/71/01/2005), pp.1-62

MoEF. 2009. Indian Forest and Tree Cover: Contribution as carbon sink. Ministry of Environment andForests, GOI, New Delhi, 8p.

Hydrological Services and Possibilities of PES from Kodagu Landscape

Kushalappa, C.G., Raghu, H.B. and C. P. Muthanna

Freshwater is a finite resource necessary for sustainable development, economic growth and political and social stability. The production of surface water is an ecosystem service that is generally not valued and paid, and hence water tariffs usually account for the services of capturing, treating, and delivering water but not for producing the water. Therefore, the goods and services provided by healthy watersheds are of critical importance to water consumers. There are many initiatives that have demonstrated that healthy watersheds provide numerous economically important services to society. In the recent years there are many examples where payments for hydrological services are being provided for upstream farmers by downstream water users. In India, Palampur Water Governance Initiative in Himachal Pradesh and initiatives undertaken by cities of Bhopal and Chandigarh are examples where such payments are in operation (Agarwal *et al.*, 2007).

Contribution of Kodagu landscape: Kodagu district is the source of origin of Cauvery, one of the important rivers in South India and hence hydrological service is the key ecosystem service of the landscape. It is therefore important to protect and preserve the Kodagu landscape in the national interest. The farming community living along the banks and majority of population living in the cities of Bangalore and Mysore are dependent on the river for their livelihood, life and prosperity. The five year average annual inflow into KRS dam in downstream for the period between 1990-95 and 1996-2000 is

186.78 TMCft and 119.65 TMCft, respectively. The water outflow from KRS also follows a similar pattern for the period i.e. 1990-95 is 71.34 TMCft and 1996-2000 is 62.73 TMCft, . These figures clearly show a reducing trend in both the inflow and outflow of water from KRS dam over the years. The sharing of this scarce water resource is a major issue of conflict between the riparian states, and Cauvery Water Tribunal established by the Central government is involved in allocating the water resources. There is also increase in the water demand for agriculture and domestic consumption within the district mainly during summer months which will impact the dry season flow. The loss of tree cover in the catchment area, land use changes, reduction in area under paddy cultivation and unregulated activities along the river banks is resulting in reduced water yields leading to major crisis of sharing the limited water not only between the states but also districts with in the state.

Rainfall distribution: Studies undertaken in CAFNET project (http://www.ifpindia.org/Managing-Biodiversity-in-Mountain-Landscapes.html)) was able to gather, analyse daily rainfall data and produce maps based on the data collected by 80 farmers over last 70 years . The analysis indicates that there is a very strong annual rainfall gradient in less than 50 kms in the Cauvery watershed.

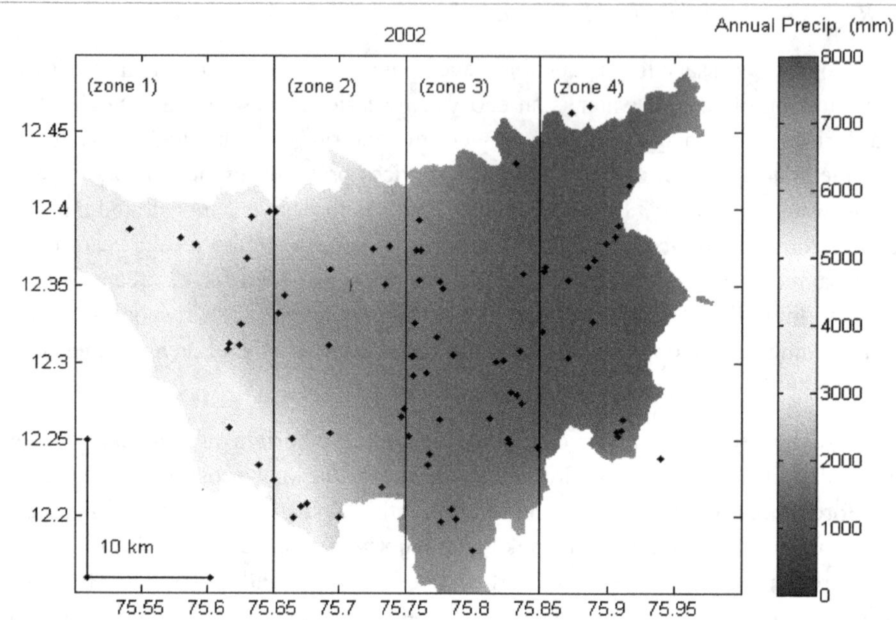

Figure 1 : Geographic position of the farms where rainfall data were collected by farmers and map of rainfall distribution generated with these data .

The analysis also showed that there is a strong fluctuation of annual rainfall with an apparent cycle of 12 to 14 years in all the four zones. From this apparent cycle, it can be predicted that the rainfall is likely to be lower in the coming years to a very low level in 2014 to 2016. Analysis also indicated that the average length of the rainy season has been decreasing over the last 35 years (1975-2010) at the rate of 0.4 day per year, and hence the rainy season has shortened on an average by 14 days over the last 35 years. Hence it is evident that the quantum of rainfall has not decreased drastically as perceived by people but there is a change in the duration of rains and the reduced dry season inflow which could be attributed to changes in the landscape due to land use changes and also increased usage of this limited water.

Role of coffee plantations in regulating hydrological services: Preliminary studies have indicated that increasing proportion of exotics in the shade cover composition had little impact on rainfall interception since trees intercept less (1-6%) than coffee plants (9 -22%). Though there is less amount of water from native plots going to rivers than from the exotic plots, the contribution of higher evapo-transpired water from native plots will have profound influence on microclimate. Further, large canopy and deep rooted systems of native species will help in percolation of water to deeper aquifers mainly during the windy monsoon periods.

Hydrological services from paddy fields: Kodagu district is an important paddy growing area (about 35000 hectares) using monsoon rains. It is estimated that about 50 lakh cubic meters of water is impounded for a period of six months in paddy fields considering that these fields hold on an average of

20 cm depth of water, which contributes to underground water recharge and hydrological cycle. However, in recent years, there is a continuous decrease in area under paddy cultivation in the region due to lower returns, and conversion of these lands to residential areas or alternate land uses. If such trends are continued, it will definitely impact the hydrological cycle mainly through reduced water recharge, which in turn affects the inflow to rivers.

Ground water management strategy: Since the terrain is undulating and receives high rainfall, major quantity of water leaves the area as run-off causing floods and heavy soil erosion. Hence, there is an urgent need to take up appropriate soil and water conservation measures in the catchment area. Studies undertaken in National Watershed Development programme in the Cauvery watershed indicates that the interventions have yielded more positive changes with respect to improved landcover contributing to improved ground water and crop productivity (RRSSC, 2005).

Valuation and payment mechanisms for hydrological services: PES for water in general is a sensitive issue, at the same time it cannot be discounted and ignored. Considering the growing demand for water for domestic and urban use, and growing Metropolitan like Bangalore, the importance of use value of water is increasing. Though the residents are paying one of the highest prices for water in the country, citizens need to be disciplined with regard to water use, which is possible with PES. Considering the hydrological services provided by landscapes of Kodagu, it is proposed that the PES for water used for non-domestic purposes should at least be 25 per cent and 10 per cent for domestic users over the existing price of water. The PES should be exempted for slum dwellers and poor population and other equity deserving segments.

The current water rate charged is Rs. 100 per acre of paddy, Rs. 400 per acre of sugarcane and Rs. 66 per acre of semi dry crops. However, even these are not being paid or recovered from farmers. Currently the estimated use value of water for irrigation is at least Rs. 1000 per acre. If the non use values and non consumptive use values are added, this value may double. Thus, PES for irrigation water should be at least Rs. 1000 per acre of irrigation irrespective of the crop cultivated at least in the head and middle reaches, since water does not efficiently reach the tail end areas.

Who has to pay for water? The user has to pay for water. In this case the user is a farmer, and due to political economy, farmers will not be/cannot be asked to pay for water. Therefore the rice mills which are processing paddy and sugar mills which are processing sugar, need to pay for water which is passed on to consumer.

Payment vehicle: Payment vehicle is the mode through which the PES payment is made. i.e. whether the user pays the PES as tax, or Cess, or duty, or advalorem tax... or fee. Since paddy and sugarcane are the most commonly cultivated crops using water, one suggestion is levying a processing fee or processing Cess per quintal of paddy and per ton of sugarcane. The PES can be five percent of the Cess on Actual Cost of Production (CACP) price of paddy processed (or Rs. 50 per quintal of paddy processed) and one percent of the minimum support price of sugarcane (or Rs. 17 per ton of sugarcane crushed). These have to be paid by the mills and then transferred onto service providers in the Western Ghats. The advantage of this payment vehicle is that the Cess can be collected and transferred on to consumers at a lower transaction cost.

Conclusions: The district is one of the well wooded regions in the Western Ghats and with sloping terrain and high rainfall, vegetation plays a key role in intercepting and regulating the flow of water. The land use and landscape changes in the form of reduction in density and diversity of trees both in natural forests and coffee based agroforestry systems, and reduction in area under paddy cultivation and conversion of paddy lands to other land use and urbanisation will result in reduced ground water recharge and lesser summer inflows into the river Cauvery. This reduced water inflow will have critical impact on meeting the growing demands of agriculture and drinking water in some of the fastest growing urban areas in Asia like the city of Bangalore. Hence there is a very urgent need to undertake research activities related to the impact of land use changes on the hydrological services and look at providing incentives to stakeholders in the upstream who are managing the landscape in a sustainable manner.

Literature cited:

Agarwal, C, C., Tiwari, M., Borgoyary, A., Acharya and Morrison, E. 2007. Fair Deal for Watershed Services in India", *Natural Resources Issues No 10* (London: International Institute for Environment and Development).

CAFNET India Final report, 2011. (http://www.ifpindia.org/Managing-Biodiversity-in-Mountain-Landscapes.html)

RRSSC/ ISRO, BANGALORE & RFSD, MOA, GOVERNMENT OF INDIA. 2005. Brief highlights of impact assessment of NWDPRA Cauvery Watershed, Kodagu District, Karnataka.

Recommendations for making PES a reality- stakeholders' perception

General recommendations:

1. There was a general consensus among stakeholders and local people representatives to look at possibilities of providing Incentive Based Mechanisms like PES for sustainable management of natural resources and to halt further degradation of landscape. The whole process can be scaled up to cover entire Western Ghats, once this model becomes successful in the Kodagu landscape.

2. Currently in India there are existing policies such as Green India Mission, CAMPA and National Forest Commission report (2006) recommendations and the draft National Water Policy (2012) which have already proposed Incentive Based Mechanisms.

3. Stakeholders felt that a strong monitoring and measuring of ecosystem services and PES mechanisms is essential. It is very important that systems need to be developed for monitoring the additionality and ecosystem service delivery for continued provision of incentives.

4. District level body for fund collection, disbursement and monitoring of activities need to be established. This body will receive funds from state, central government, corporate sector and international organizations. The fund should be exclusively utilized in such a way that the activities taken should foster sustainable utilization of resources and development of livelihoods. This body will also pay for undertaking research and cost of monitoring PES activities in the district.

5. Rather than valuing and incentivising individual services, recent concepts like landscape labeling which considers bundling of services and cultural attributes of the landscape as defined by local communities could be considered.

6. There is a need to create awareness on PES among the different stake holders.

7. Driven by economic gains, agricultural lands have been converted to non-agricultural purposes which need to be regulated through appropriate legal mechanisms.

Specific Recommendations:

Biodiversity and Carbon:

1. Diversion of forest lands to non forestry activities have occurred in the past. Extent of such lost forest areas has to be assessed and valued, and must be considered while developing incentive mechanism for the entire district.

2. State Forest Department has to exercise strict vigilance in sensitive areas along the district border by deploying more staff and establishment of anti-poaching camps in order to restrict illegal activities.

3. The diversity of natural forests needs to be improved by restoration of monoculture teak plantations, enrichment of degraded natural forest areas and management of invasive weeds.

4. Government driven conservation efforts aiming at acquisition of biodiversity rich areas (Jamma malais and Coffee Saguvali malais), revenue forests around protected areas need to be taken up with suitable compensation measures.

5. There is a need to survey and demarcate sacred groves and devise appropriate incentive mechanisms based on joint forest management systems.

6. The density and diversity of native tree cover needs to be maintained and sustained based on recommendations of Coffee Board and CAFNET project results. In this regard, appropriate incentive mechanisms like eco and forest certification need to be promoted.

7. Providing Elephant-friendly coffee label (Aane kaapi, in Kannada) could help raise the tolerance level within the local population. If farmers obtain a tangible benefit from the presence of elephants in their estates, the problem may cease to be perceived so negatively.

8. Studies undertaken have revealed that Silver oak should not exceed more than 30 % of the shade trees in coffee agroforests to sustain biodiversity and carbon sequestration without affecting the coffee productivity. Awareness to promote planting of native species have to be done in the interest of maintaining the flow of ecosystem services from the landscape.

9. Exotics like Silver oak are becoming more popular due to the tree ownership issues and fast growth rates. Hence in order to maintain the native trees and to take up enrichment planting of these trees, tree rights should be given to land owners and their harvest be regulated as per the existing rules under Karnataka Tree Preservation Act of 1972.

10. There is a lack of availability of seedlings of native tree species. Therefore, strengthening of forest department nurseries in order to raise sufficient quantity and quality seedlings of native species, with public-private partnership should be encouraged.

11. An area with research significance or potential for bio-prospecting can be made accessible to research agencies or pharmaceutical companies on an agreed payment regime. With proper surveillance of, and association with forest department and local bodies, this would entail the rights of collection, testing and usage of genetic material for either research or product development.

12. Payments for Agrobiodiversity Conservation Services (PACS), a global initiative from Bioversity International need to be explored to encourage individuals and communities (http:// www.bioversityinternational.org/research/sustainable_agriculture/pacs.html) involved in this endeavor. This is particularly important in Kodagu as traditional paddy farming with local varieties is still in practice.

13. As regards to taxes as payment vehicle for carbon, it is envisaged to collect 5 per cent of the cost of vehicle as Green Tax as one time life time tax during registration. Another five per cent of the cost of a vehicle towards biodiversity tax is recommended. This can be collected by the respective Regional Transport Offices.

Water:

1. There is an urgent need for an action plan to protect and promote paddy cultivation. This could be done by regulation in line with the law of Kerala Government and endorsement by Zilla Panchayat of adjoining Dakshin Kannada district. Incentives for paddy cultivation like providing Rs.5000 per acre as in Kerala or better support price covering the cost of Rs. 2500-3000 per quintal is essential.

2. There is a need to enforce the existing rules of river bank protection laws to prevent illegal encroachments, sand mining and pollution.

3. Creation of Cauvery Watershed Development Authority in line with Command Area Development Authority (CADA) comprising of stakeholders from government, farmers, water users, institutions and NGOs to prepare and implement action plan for watershed development in the catchment area.

4. Part of the tax collected by Cauvery Niravari Nigama and CADA should be provided to the Kodagu fortaking up watershed development activities.

5. The PES for water is estimated at around 5 per cent of the Cess on Actual Cost of Production (CACP) price of paddy and 1 per cent of the Minimum Support Price of sugarcane to be obtained from processors of paddy (the rice mills) and of sugarcane (sugar mills)

Tourism:

1. There is an urgent need to regulate tourism in the district. This can be done by reviving and strengthening the District Tourism Promotion Council (DTPC) by inviting stakeholders to be part of this committee. This body should take up carrying capacity study on tourism and till such time there should be a moratorium on issuing new license for resorts and big hotels.

2. There is a need to promote locally managed ecotourism ventures like Home Stays which has little impact on the environment. All the home stays should register with Coorg Tourism Promotion Body, a private consortium of home stay owners meant to regulate home stays and their activities.

3. Local bodies like Town Municipalities and Gram Panchayats should not only regulate issuing of new licenses for tourism ventures but also collect one time tax from the applicants and annual renewal charges which could be used for development of sustainable natural resources management.

4. There is an urgent need to shift the tourism zone in Rajeev Gandhi National park from the middle tothe periphery to mitigate disturbance to wildlife.

Action forward

Based on discussions with stakeholders and ongoing initiatives under Model Forest Programme, following two models will be attempted as pilot studies.

Model – I: CSR based approach

Using contributions from private entrepreneurs under corporate social responsibilities, the ongoing green village community forums established under Model Forest Programme activities in three villages will be strengthened and extended to additional villages. The private entrepreneurs could be local corporate in plantation sector like TATA/BBTC/Skanda Coffee or from tourism sector.

Model – II: Upstream – Downstream resource sharing approach

Efforts will be undertaken to formulate a payment mechanism for hydrological services by water users in Madikeri town and service providers in the upstream.   The town municipality can collect water Cess from commercial and domestic users and provide incentives to farmers. Experiences from the towns of Bhopal, Chandighar and Palampur will be employed in developing this model.

Kodagu district is one of the greenest landscapes in India and is part of the Western Ghats, a global hotspot of biodiversity. However, the landscape and demography in the district is currently undergoing rapid changes.

There is an urgent need to formulate an action plan for economic development model based on sustainable utilization of natural resources involving Incentive Based Mechanisms like Payment for Ecosystem Services (PES), Ecological and Forest Certification, and Landscape Labeling. Hence an effort to review the key ecosystem services from the landscape and mechanisms of providing incentives to communities has been accomplished in this document.

RECOMMENDATIONS OF THE COORG WILDLIFE SOCIETY ON THE IMPLEMENTATION OF THE KASTURIRANGAN[HLWG] REPORT IN KODAGU

1. Before making any recommendations on the implementations of the Kasturiragan report in Kodagu, it will be essential to study certain key aspects of the report and its implications in the context of the Kodagu Landscape. There is a need to focus both on the positive aspects of the report and also on the various anomalies.

2. In the Preface of the report it is stated that the report of the Western Ghat Ecology Expert Panel [WGEEP report] together with the Kasturirangan report can be a starting point in understanding the man-environmental relations in the Western Ghats. Here it is pertinent to note that the WGEEP report had recommended all the three taluks in Kodagu as Eco Sensitive Zone-1, whereas, the Kasturirangan report recommends only 53 villages in Kodagu as Eco Sensitive Areas [ESA].

3. In Chapter 1, Introduction the opening paragraph states as follows:

   The Western Ghats is the home for about 50 million people belonging to the Six States of the Country. It is the source of water for the entire Peninsular India, and also influences the monsoons. The life supporting and biodiversity rich ecosystems of Western Ghats are threatened today due to habitat pressures.

   It is therefore crucial to identify ecological importance of specific areas of the Western Ghats with special reference to hydrological services. The focus of the recommendations of the Coorg Wildlife Society is on the very crucial aspect of identifying the importance of Kodagu as the Catchment of River Cauvery and the specific threats, challenges and opportunities with regard to the Kodagu landscape.

4. Chapter 3 of the report deals with the impact of Climate Change on the Ecology of the Western Ghats. Paragraph 3.4 reads as under:

   It is also seen that there is an increase in the moderate drought development for Krishna, Pennar, and Cauvery basins, which have either predicted decrease in precipitation or have enhanced level of evapo-transpiration

   This is an important statement since it underlines the problem of water security for the Cauvery fed regions in the future. Kodagu, being the Principal Catchment of Cauvery River will therefore need enhanced protection for its eco systems.

5. Under Chapter 5, Paragraph 5.2.1. the report states:

*The Western Ghats form the major watershed in Peninsular India and as many as
Fifty Eight major Peninsular Indian rivers originate from it.*

In the case of Kodagu, the Kasturirangan report [HLWG report] has recommended a
number of villages along the Western Ghat Ever-Green forests as ESAs. Three of the
major tributaries of River Cauvery originate in these ever-green forests. The
Kasturirangan report makes a mention of watershed, but in the case of Kodagu, the
protection is only for the areas of origin of the rivers. This does not serve much purpose.
The entire area of Kodagu needs to be identified as a single eco-region and as a principle
catchment and watershed of Cauvery

6. <u>Population density</u>

Under paragaraph 5.4.3.1 of the report, it is mentioned that population density is one of
the criteria for identification of ESAs. The table shows an average population density of
ESA villages is 86.96 for ESA villages and 411.91 for the Non-ESA villages. It is for
consideration that Kodagu has a very low population density except for Madikeri City
Municipality and a few large towns. Therefore, the population density for Kodagu needs
to be reassessed in the context of ESA.

7. **<u>WGEEP recommendations for sector level planning and their implications</u>**

In Chapter 6 of the report, Para 6.1, states:

*The area defined as 'cultural' has been deliberately identified and segregated from
the 'natural' landscape. This does not mean that these settlements, plantations or
agricultural fields do not co-exist on the biological diversity of the natural area or*

*that these areas have an open license to pollute or degrade the environment. It is for.this
reason that HLWG proposed to recommend a higher level of scrutiny and*

monitoring for projects within 10 km of the ESA.

*HLWG also recognizes that this cultural landscape is biologically rich. For instance,
the coffee plantation areas of Kodagu have high biological diversity in the cultural
landscape. The sacred groves of many settlements are scattered and so not
detectable through remote sensing. But these groves are the most abiding symbols*

*of people's belief in the protection of nature. HLWG has recommended policies to
incentivize these practices so that growth across the Western Ghats can be
environmentally sound.*

The HLWG recognizes the importance of the area identified as cultural landscape in
Kodagu, but has failed to make a mention of the paddy fields that constitute about 10% of
the landscape and provide valuable hydrological services by acting as wet lands. The
report does make a mention of plantation areas of Kodagu and their high biological value.
But the recommendations of the report will not prevent the rapid urbanization of Kodagu

and steady loss of these paddy fields and Coffee plantations for construction of house-sites, layouts and resorts.

8. In Chapter 6, Paragraph 6.2, The HLWG has rightly pointed out the implications of the recommendations of the WGEEP and has made a specific mention of the Coffee Plantations of Kodagu. However, the fact of the matter is that the WGEEP Recommendations provide for protecting these plantations as well as wet lands from further degradation due to land conversion and urbanization. Specific clauses of the WGEEP report could have been reviewed and revised in detail in consultation with the concerned stake holders. ***There was no need to throw out the Baby with the Bath water!***

9. <u>6.4.3 Power/Energy, including hydropower and wind</u>

In Chapter 6, paragaraph 6.4.3, hydro electric projects have been permitted subject to certain conditions. It is for consideration that there is scope in Kodagu, for a number of hydro-electric projects to come up in the evergreen Western Ghat forests. These forests are rich in faunal biodiversity and backwaters of the hydro electric projects, however small, fragment the forests and restrict movement of wildlife. In Kodagu, this will lead to increased level of Human Elephant Conflict. The Hydro Electric Project in Karike area of Kodagu, in the evergreen hill forests is a case in point. An area of 3.144 hectare of evergreen forests in the Pattighat Range was diverted for the project. The related infrastructure such as construction for office and accommodations for staff also affects the adjoining areas. The Elephant problem in the adjoining areas has become more severe since the completion of this project. Even if there is a minimum distance of 3 Kms between the projects, as recommended, the collective damage would be substantial.

10. <u>Settlements</u>

The recommendations made in Chapter 6, paragraph 6.4.5 are perhaps the most damaging. It implies that projects up to 20,000 sq meters are permitted even in the designated ESAs. Therefore, for example, ten different projects of an average of 19 000 sq meters but each less than 20,000 sq meters will be permitted. That totals to 1,90,000 square meters! Unless there is are strict restrictions on any land conversion for non agriculture or non plantation commercial purpose, these recommendations become meaningless and will in-fact only serve to hasten the process of destruction of the ESAs.

11. <u>Infrastructure including Transport</u>

Kodagu is a tiny district of 4108 sq kilometers. In addition to the trend for rapid urbanization, the landscape is set to be ripped apart by a slew of projects including Power Lines, Highways and Railway Lines. All these projects are for the benefit of other areas adjoining Kodagu, and could well be avoided. The 400 KV Mysore-Khozikode High tension Power Line through Kodagu destroyed thousands of trees in the private coffee plantations of Kodagu, causing severe ecological damage and has increased the level of

Human- Elephant conflict that has already claimed a number of lives of the local communities.

It is for consideration that the recommendations of the HLWG do not make any meaningful provision for the protection of the ESAs from such projects that are devastating for the Environment of Kodagu.

12. Tourism

With reference to paragraph 6.5,tourism has done tremendous damage to Kodagu and the term eco-tourism exists only on paper. Resorts are mushrooming in Kodagu and there has to be a moratorium on issuing any fresh licenses for resorts and other tourist accommodation and for expansion of existing facilities until a proper evaluation is carried out on the carrying capacity for tourism in Kodagu. In the long term, Kodagu will suffer if a 'tourist-dependent economy' is encouraged.

13. Incentivizing green growth in Western Ghats

In paragraph 6.5 of Chapter 6 of the report of the HLWG, there are excellent recommendations on incentives for Green growth. In this regard, Payment for Ecological Services is a vital component of the incentives. The College of Forestry, Ponnampet, has prepared document of the Scope for Payment for Ecological Services in Koagu. This must be taken forward and implemented in Kodagu as a pilot project for the entire Western Ghat region.

14. Water for Mumbai

In paragraph 6.5 of Chapter 6, there is a mention of water supply to Mumbai:

"For instance, the city of Mumbai, gets its water supply from the forested watersheds located over 100-110 km away. The city, which is already water stressed, will be in dire straits, if the forests of Western Ghats, are not protected or regenerated. Currently, the city also does not pay for the ecological cost of conservation of the Forests".

This is an excellent observation by the HLWG. Kodagu provides food and water security for millions of people across South India. It is not only the forests but also the coffee plantations and paddy fields that serve as the Principal catchment and Watershed for Cauvery River. It the Kodagu Landscape is not protected, it will have serious impacts on the economy and the food and water security of South India. ***It is therefore in the National Interest to protect and preserve the Kodagu landscape that is a Principal Catchment of River Cauvery, providing over 70% of the total inflow into the KRS Dam.*** Against this back ground it is a matter of serious concern that the critical ecological balance in Kodagu is under severe stress due to a variety of reasons. This has manifested in reduced inflow into the KRS Dam over the past ten years. While there is constant tussle and conflict between Karnataka and Tamil Nadu on sharing Cauvery water, little thought is given to the relentless degradation of the Principal Catchment Area! The Mumbai example also makes a strong case for payment for ecological services for Kodagu.

15. <u>Recommendations of the Coorg Wildlife Society</u>

In the context of these observations, the recommendations of the Coorg Wildlife Society are as under:

[a] The measures for protection of the Kodagu Landscape should be a mix of the HLWG recommendations and the WGEEP report. All the three taluks of Kodagu should be declared as Eco Sensitive Zone as recommended by the WGEEP. Only this can ensure the protection of the Principal Catchment and Watershed of River Cauvery in its totality. However, as mentioned in the report of the HLWG, some of these recommendations on Sector Level Planning could have adverse implications on the livelihood and social security of the local communities. Therefore these recommendations of Sector Level Planning will need to be implemented only after being reviewed in consultation with the concerned stake holder groups. The Coorg Wildlife Society had made recommendations to the HLWG regarding the WGEEP report. A copy of the recommendations made by the Coorg Wildlife Society is attached

[b] <u>Urbanization and Land Conversion</u>

Urbanization and land conversion need to be curtailed on an urgent basis. This is the single greatest threat to the Kodagu Landscape. There should be strict regulations on conversion of agriculture or plantation land for commercial non-plantation or non-agriculture purposes such as resorts, layouts, house sites etc.

[c] <u>Town delimitation</u>

Townships are expanding without any planning. There should be strict town delimitation. Buildings must be restricted to a maximum of Ground floor plus one floor. This is being implemented in some of the ESAs in India that have been ratified under the provisions of the Environment Protection Act.

[d] <u>Tourism</u>

Kodagu is being subjected to invasive tourism and proliferation of resorts. This is causing severe damage to the serene environment of Kodagu and to the fragile socio-economic fabric of the local communities. There is a need for a moratorium on resorts and other tourist accommodation and for expansion of existing facilities till a study is taken to analyze the carrying capacity of Kodagu for tourism.

[d] <u>Infrastructure projects</u>

There are a number of infrastructure projects that are to pass through Kodagu such as Railway lines, highways and Power Transmission lines, mainly for the benefit of neighboring areas. Such projects should not be permitted and alternate alignments should be worked out.

[e] <u>Payment for Ecological Services</u>

The document prepared by the College of Forestry, Ponnampet needs to be evaluated by policy makers. The concept should be applied to Kodagu as a means of providing financial and social security to the communities of Kodagu. A copy of the document is submitted herewith for your perusal.

16. In conclusion, we have made our recommendations in the hope that it will pave the way for a Win- Win situation in declaring the entire area of Kodagu as an Eco-Sensitive Zone. These recommendations will ensure the water security for the millions that depend on the Cauvery Waters and will provide for the economic development of South India. At the same time it will cater for the well being of the Kodagu People.

Col CP Muthanna
Coorg Wildlife Society

HABITAT MANAGEMENT IN THE NILGIRIS BIOSPHERE RESERVE ANDTHE ELEPHANT RESERVES OF SOUTH INDIA

NILGIRIS BIOSPHERE RESERVE AND ADJOINING FORESTSNILGIRIS BIOSPHERE RESERVE-5500 SQ KMS &TOTAL AREA IS 12,587 SQ KMS

ELEPHANT POPULATION APPROXIMATELY 10,000

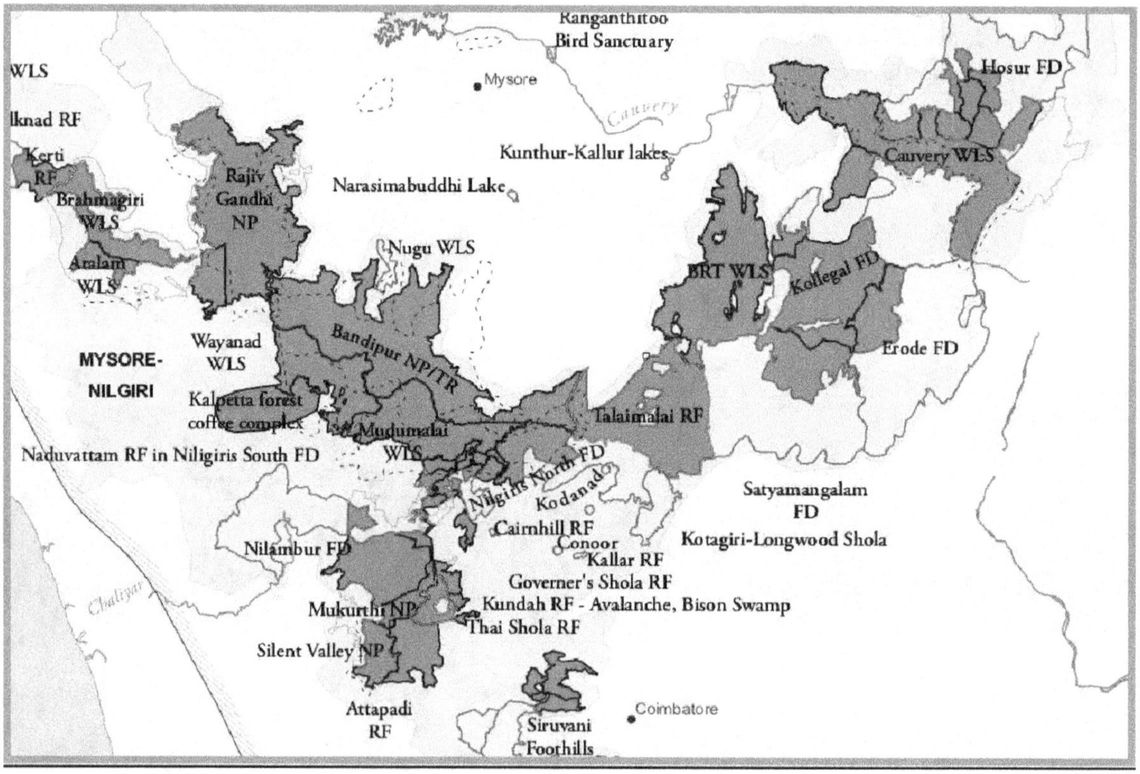

<u>PREAMBLE</u>

During the past decade there has been a sharp escalation in incidents and frequency of Human Animal Conflict, involving Elephants, Bears, Gaur, Leopards and Tigers in South India. Let us look primarily at Human Elephant Conflict [HEC] since this has been a cause of huge concern. While most discussions are around management of the species, i.e trans-location, setting up barriers, and population management, this paper does not attempt to make a case against species management. However, it is for consideration that in the cacophony of discussions, debates and opinions on species management, we have lost sight of the fact that the elephant is in many ways acting as a messenger telling us that there are serious problems in the habitat.

The habitat in South India is in essence the Nilgiris Biosphere Reserve [NBR] and adjoining Elephant Reserves spanning over 12000 sq km in the three states of Karnataka, Tamil Nadu and Kerala. The elephant reserves are namely the Mysore Elephant Reserve, Wayanad Elephant Reserve and the Nilgiris Elephant Reserve. This is perhaps one of the largest network of Elephant habitat in South Asia. It is also home to the largest population of wild Asian Elephants with an estimated number of about 10000 elephants. The degradation of habitat is due to several reasons including teak monoculture, proliferation of invasive species, forest fires, and loss of forest cover due to development projects. Over the years, the problem has only increased because the basic concept has been to 'Shoot the Messenger' rather than to heed the message.

On the other hand, if we fail to act and if there is continued degradation of this forest landscape, wildlife species will increasingly fall prey to inbreeding and genetic disorders that will degrade the species within a few generations. This in turn will again contribute to forest degradation. Isolated interventions in scattered pockets will not yield long term results and therefore we must look at the entire landscape covering the elephant range in South India.

<u>PROPOSAL FOR HABITAT MANAGEMENT IN THE NILGIRIS BIOSPHERE RESERVE ANDTHE ELEPHANT RESERVES OF SOUTH INDIA</u>

<u>Introduction</u>

The Nilgiris Biosphere Reserve [NBR] and adjoining forest areas of over 12000sq kms across the three states of Karnataka, Kerala and Tamil Nadu India is one of the largest networks of Elephant Habitat in South Asia. Apart from immense value as a prime habitat for wildlife it also provides vital hydrological services to South India in terms of Watersheds and Catchment areas. Moreover, this forest landscape should also be viewed as a vitally important component in providing resilience to the impending effects of climate change in this region of South India, which is home to several million people.

Against this background, it is a matter of serious concern that the sharp escalation of Human- Animal Conflict and especially Human-Elephant conflict in areas adjoining the NBR and the Elephant Reserves in South India indicates high levels of habitat degradation in this vitally important network of PAs and Reserve forests. A 'business as usual' attitude will lead to serious consequences in the coming decades. It is therefore extremely important to come to grips with the problems and challenges in Habitat Management in the NBR.

<u>Major Problems and Challenges in the NBR and adjoining elephant reserves can be summarized as below</u> under Forest Degradation and Disruption of Elephant corridors and Migration routes

A. <u>Forest Degradation</u>

(a) <u>Teak Monoculture</u>-vast portions of the NBR and adjoining area are under teak monoculture which have changed the original forest regime and adversely affected the microclimate. Teak also deprives fodder for elephants though elephants have now been compelled to feed on Teak bark. Teak restricts natural regeneration of other indigenous tree species. There is a need to phase out teak monoculture and this should to be permitted in PAs in the NBR.

(b) <u>Invasive Weeds</u>- Massive proliferation of invasive weed species such as Senna Spectabilis, Lantana Camera and Eupatorium are principal causes of forest degradation resulting in a stifling of indigenous tree and plant species and restricts the fodder available for elephants and other wildlife species.

(c) <u>Forest Fires</u>- Large scale forest fires occur frequently in the dry season and destroy large areas of forests. Forest fires also encourage the proliferation of invasive weed species. There is a need to establish a proper integrated system for prevention and tackling forest fires. Available expertise could also be taken from other countries such as Canada where the Canadian Forest Service provides training in setting up early warning systems. Technological systems such as the Fire Detection Robot introduced by other countries could also be developed in India would also be very useful.

(d) <u>Employment of Ecological Territorial Army Units:</u> These units are formed with a core staff of serving personnel and the bulk of the manpower is from able bodied retired army personnel. There are about 8 such units in different states of India and they have done excellent work in forest land restoration. These could also be deployed in the South Indian states for forest land restoration as well as for prevention of forest fires

(e) <u>Grazing pressures-</u> Grazing pressures by livestock prevents natural forest rejuvenation and is a potential threat to herbivores through spread of diseases such as foot and mouth disease that could wipe out entire herds of wildlife. The local communities need to be provided alternative grazing areas and training and capacity building for stall fed cattle and Diary Cooperatives. There is also scope for cultivation of millets etc in fallow paddy lands for use as cattle fodder. The project by Anna Hazare Model in the Ahmednagar region of Maharashtra is an excellent success story in stall fed cattle and diary cooperatives for rural areas.

B. <u>Disruption of Elephant corridors and Migration routes</u>

It is important to understand that when we improve the migration corridors and quality habitat in the Nilgiris Biosphere reserve, and the Elephant Reserves, it will benefit all forms of wildlife, improve water sheds & catchment areas and provide the region with greater resilience from the effects of climate change. Fragmentation of habitat and disruption of elephant corridors and migration routes will have severe and irreversible damage to genetic health of all forms of wildlife and will in turn contribute to degradation of habitat. The major causes of disruption of corridors and migration routes are:

(a) <u>Forest Land encroachment</u>

Improper implementation of the Forest Dwellers Act has resulted in large settlements springing in forest areas, blocking elephant movement and forcing the elephants to move outside the forests. Critical migration routes have also been cut off due to projects such as the Kabini Dam. While it is a politically sensitive issue, this trend must be checked before the situation crosses the point of no return. The improper implementation of the Forest Dwellers Act is also a major problem.

(b) <u>Linear intrusions and development projects</u>

Large swathes of forests are lost due to Forest Conversion for linear projects. The issue of forest land diversion in critical wildlife habitat for roads, railways lines and power lines has led to further degradation and blocking of elephant corridors. The background paper for the National Board for Wildlife on 'Framing ecologically sound policy on linear intrusions affecting wildlife habitats' prepared by the Nature Conservation Foundation in Jan 2011 needs to be ratified. **It would be extremely important to understand that while taking steps for improving the habitat, further degradation of the habitat by means of linear development projects and dams etc that require further**

diversion ofthese forest areas should be avoided.

(c) Blocking of Migration routes

Migration routes have been blocked due to projects such as the Kabini Dam. There is a need to analyze the feasibility of restoring these routes at least partially through engineered structures. For example, over four hundred overhead wildlife crossing places have been built across a highway in the Netherlands.

(d) Campa Funds

There is a need to stop Campa funds being used for any monoculture plantations. On the other hand, Campa funds should be utilized for conserving diversity of over 200 floral species.

(f) Tourism

Excessive tourist pressure in protected areas cause disturbance to wildlife and drive elephants out of forest areas. This is another cause of HEC. Movement of tourists in PAs must be regulated to the extent possible.

(g) Elephant population

There is one aspect regarding elephant population that needs consideration. While some experts state that there is no major increase in the overall population of elephants, there is also a growing perception that the population of elephants in this region has increased considerably over the past few decades. There have been some suggestions regarding sterilization of female elephants, but the practicability of such measures will need to be validated.

PROJECT PROPOSAL

A project proposal has to be prepared for addressing all the issues as mentioned above.There must be a time bound program for:

(a) Preparing the proposal and obtaining funding and other resources

(b) Capacity Building

(c) Project implementation

After the Capacity Building phase, the project implementation phase should be completed in five years. The Forest Restoration program will need to continue over a longer period. This could be reviewed during the third year of the implementation phase.

<u>PROJECT PARTNERS AND STAKEHOLDERS</u>

The project for management of the Nilgiris Biosphere Reserve and the three elephant reserves will require the collaboration at the global, state and regional levels.

(i)   **<u>Global level</u>**- Global Partnership for Forest Land Restoration, Bonn Challenge, Forest Land Restoration such as RECOFTC [Thailand], National/ Central level Project partners and stake holders including MOEF India, WTI, WWF, IUCN India, IUCN Commission for Protected Areas Wildlife Institute of India,

(ii)   **<u>State level</u>**- Centre for Ecological Sciences [IISC Bangalore], IUCN India, Forest Departments of Karnataka, Kerala & Tamil Nadu. Perhaps WWF, could set up a coordination centre at a suitable location.

(iii)   **<u>Regional level</u>** - Coorg Wildlife Society, College of Forestry Ponampet, Environment and Health Foundation, Kodagu Model Forest, Forest First Samithi etc.,

<u>Conclusion</u>

It is obvious that urgent and collaborative action is required to implement measures to restore and preserve the NBR and adjoining Elephant Reserves. Procrastination will only increase the degree of difficulty. On the other hand, a coordinated effort by all concerned agencies offers hope for the future of this precious forest landscape. The success of this program will also serve as a model for other degraded forest areas both in India and otherregions of the world.

Col CP Muthanna
Former President,
Coorg Wildlife Society
& Vice Chair, Kodagu Model Forest Trust
Founder and Hon Secretary,
Environment and Health Foundation [India]

A NOTE ON PERMAFROST AND THE IMPLICATIONS OF PERMAFROST MELT IN THE HIMALAYAS

What Is Permafrost?

Permafrost is any type of ground, from soil to sediment to rock, that has been frozen continuously for a minimum of two years and as many as hundreds of thousands of years. It can extend down beneath the earth's surface from a few feet to more than a mile, covering entire regions, such as the Arctic tundra, or a single, isolated spot, such as a mountaintop of alpine permafrost. Permafrost is also present in the Himalayas

- Beneath its surface, permafrost contains large quantities of organic leftover from thousands of years prior, to include dead remains of plants, animals, and microorganisms that were frozen before they could rot.

- It also holds a massive trove of pathogens which could cause disease outbreaks or even pandemics.

- When permafrost thaws, microbes start decomposing this carbon matter, releasing greenhouse gases like methane and carbon dioxide. **However, the most alarming aspect of permafrost thawing is the fact that when permafrost melts, the trapped carbon is released into the atmosphere in the form of Carbon Dioxide and Methane, which is a powerful greenhouse gas. This sets off a vicious cycle of climate change.**

- Researchers have estimated that for every 1 degree Celsius rise in temperature, these grounds could release GHGs to the tune of 4-6 years' of emissions from coal, oil, and natural gas.

- Along with greenhouse houses, these grounds could also release ancient bacteria and viruses into the atmosphere as they unfreeze.

- The most alarming aspect of permafrost is the rapid release of massive amounts of Methane and Carbon Dioxide into the atmosphere from subsea permafrost melt. The quantification of these emissions is complex; however scientists opine that emissions from subsea permafrost melt are substantial and that it is an important factor for modeling of climate change scenarios.

Permafrost melt in the Himalayas

Permafrost melt was first reported in the Tibetan region of Qinghai during September 2017 when a thick stream of melted permafrost flowed like lava through a fertile valley in Yushu, Qinghai on September 7, destroying a mini-van and a tent.

When permafrost melts, the land above it sinks or changes shape and the shifting ground could potentially damage buildings and infrastructure such as roads, airports, and water and sewer pipes. It also causes landslides, slope collapse, and glacial lake outburst floods (GLOFs) and topples trees along the path of the floods.

Downstream areas are also vulnerable to the impacts of permafrost thawing. It is feared that Melting of Himalayan permafrost would contribute to river floods downstream in neighboring Bangladesh also. According to ICIMOD experts, glacial lakes or hydropower dams in the vicinity of mountains with permafrost will witness an increase in rock fall hazards, which may lead to outburst floods with grave downstream consequences.

For the first time, the Indian government is planning a pilot study of permafrost in Himachal Pradesh. It is expected to provide crucial data in a little-studied area of climate change impact in the Himalayas.

Permafrost melt is a global concern. In the Indian context, it needs to be understood that permafrost melt in the Himalayas will accelerate climate change in the region. This is all the more reason for efforts for mitigation of climate change in the Himalayas by reduction of Black Carbon emissions combined with large scale forest land restoration programs as elaborated in the HIMEK Alliance proposal.

Stabilization of Climate Change in the Himalayas:Strategy for a Regional Response

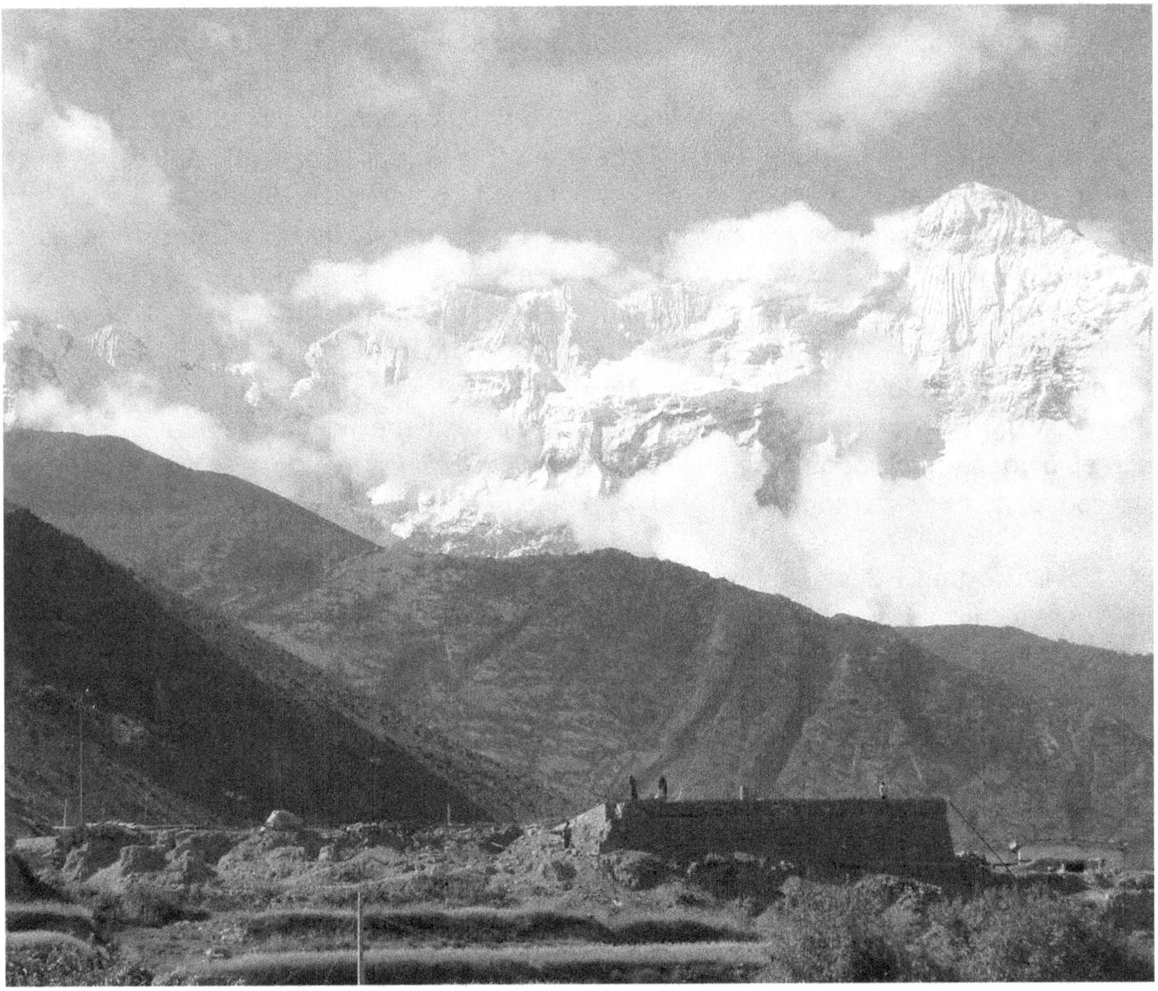

Stabilization of Climate Change in the Himalayas: Strategy for a Regional Response

Over the past two years, the Copenhagen climate meeting was widely heralded as an 'historic event' at which a new global change regime would be adopted by Heads of State and Governments of a large number of UNFCCC Parties. In the event, the results of the meetings did not live up to the expectations and they ended with an unambitious, non- legally binding Accord being 'taken note of' by the Conference of the Parties. Carbon dioxide has been the focus of these discussions, but science tells us that there are other emissions that are impacting our Earth. Black carbon (BC), commonly known as soot (air pollution), which is emitted from incomplete combustion of fossil fuel, biofuel and biomass burning, is now attributed as the second largest contributor to global warming. Other non-CO_2 greenhouse gases (GHGs), such as carbon monoxide and nitrous oxide, also contribute to global warming.

IUCN believes that BC and other non-CO_2 GHGs are a critical issue particularly in Asia and these are something that can be tackled regionally without getting tangled up in lengthy and expensive global negotiations. It would be expedient to combine the reduction of these emissions with a regional initiative to regain the mitigating effects of the Himalayan forests.

The Himalayas

The mass of ice and snow in the Himalayas is the third largest in the world after the Greenland and Antarctic ice sheets. The Himalayas are vitally important in sustaining the lives and livelihoods of millions of people. It is therefore a matter of deep concern to the entire global community and to the people of South & Southeast Asia and China in particular that the Himalayan environment is under serious threat due to the effects of climate change and global warming.

The glacier and snowmelt of the Himalayas are the source of several principal river systems in South & Southeast Asia and China: these rivers sustain one fifth of the human population. The Himalayas have a profound effect not only on the regional climate, but also on the climate of the Earth. The Himalayan ranges are also a treasure trove of biodiversity including unique alpine and temperate conifer forests and hundreds of precious medicinal plants and herbs. Apart from the direct impacts of climate change, rising temperatures could also result in proliferation of pests and invasive species. Climate

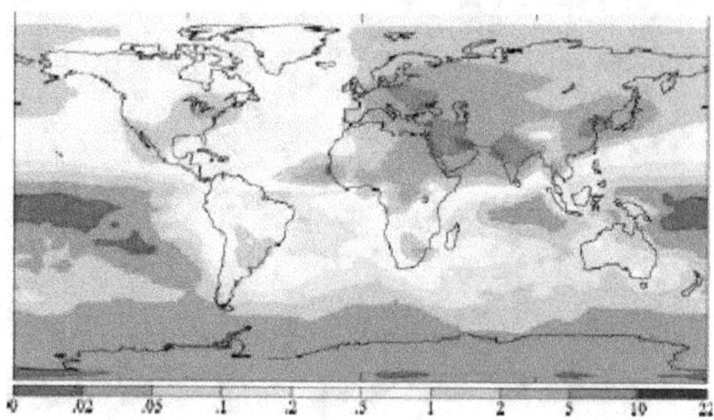

Simulated soot density for the world (Source: NASA Earth Observatory Website)

change in the Himalayas on average is at a much higher rate than the rest of the world. The rate of temperature rise from 1991 to 2007 was 0.76ºC per decade in the Himalayan region of China. On the Tibetan plateau, temperatures rose by 1ºC in the last decade alone.

The mountains and valleys of the Himalayan region are home to 100 million people that include indigenous communities whose livelihoods and culture are closely linked to the mountain ecosystems and forests. They have an uncertain future in the face of climate change. Increased temperatures will have a drastic impact on their water and food security and on forests, agriculture, and horticulture that are major sources of sustenance and income to large sections of the Himalayan population. They will become increasingly dependent on food imports and will be more vulnerable to flooding and glacial lake outburst floods (GLOFs).

Climate change has already taken a heavy toll of the Himalayas. The World Watch Institute states that due to the effects of global warming the pattern of precipitation in the Himalayas and the regions contiguous to the Himalayas will undergo a more drastic change in the years to come. The increase in temperature will reduce the amount of snowfall and therefore the snow fed rivers of China and the Indian subcontinent will have less water flow in the summer months when the snow melts. However during the monsoons the effect will be excessive rainfall, leading to more intense flooding.

The Himalayan glaciers are receding faster than anywhere else in the world. The Gangothri glacier that is the source of the Ganges has receded by 600m in the past 40 years. There has been a marked increase in the rate since 1971 and the glacier has been shrinking by 30m per year. Surveys based on satellite images and ground investigations by ISRO's Space Application Centre (SAC) Ahmedabad [India] have established that in Himachal Pradesh state of India, the glaciers have reduced from 2,077 km2 to 1,628 km2 – an overall deglaciation of 21% in four decades. According to SAC as many as 127 glaciers of less than 1 km2 have lost 38% of their geographical area since 1962. The larger glaciers, which are progressively getting fragmented, have receded by as much as 12%.

Regional causes of change

Black carbon
Black carbon (BC) is a type of aerosol, defined by its chemical reactivity and the amount of light it absorbs, and mostly consisting of soot, charcoal and other products of incomplete

burning of organic matter (IPCC). It strongly absorbs light and is thought to be accountable for a large part of global warming and for regional warming in the Himalayas and the Arctic – known as climate forcing.

Aerosols are particles suspended in air. When aerosols exist in the atmosphere they can interact with incoming solar radiation in a variety of ways, depending on the type of aerosol – i.e. its chemical composition, size and shape. Some aerosols, for example BC, absorb solar radiation, while some, for example sulphates, mostly scatter and reflect it. Both of these mechanisms reduce the amount of radiation that reaches the earth's surface, and hence reduce heating of the earth's surface. This is sometimes referred to as surface dimming.

When aerosols absorb solar radiation they heat up, which in turn heats the atmosphere in which the aerosols are suspended, often at high altitude. Hence scattering aerosols produce less atmospheric heating than absorbing aerosols, for example BC, situated in the same region of atmosphere.

Absorbing aerosols like BC also have an effect on solar radiation when they are deposited onto snow, ice, or other bright, reflective surfaces in that they affect the extent to which the surface reflects light. Reducing the reflectivity of the surface increases the direct solar heating of that surface.

Emissions in China and India are key drivers of regional climate in the Himalayas: 25-35% of global BC in the atmosphere comes from China and India. Direct and indirect BC climate forcing could accelerate the deglaciation in the Himalayas region. This could have severe consequences, as the livelihoods of millions of people depending on river systems such as Indus, Ganges, Brahmaputra, Mekong, Yangtze, etc. will be affected.

Other climate effects

Atmospheric Brown Cloud (ABC) that are formed by aerosols, including BC, also have an impact on the hydrological cycle. The cycle is driven by the solar heating of the surface which is weakened by aerosols blocking the light. Cloud droplets condense on aerosols, and high concentrations of condensation nuclei can mean drops cannot grow big enough to fall as rain or snow. These two effects could mean that BC reduces rainfall in the Himalayas on average. It should be noted that while other GHGs increase the heating of the surface and hence increase evaporation and thus rainfall, aerosols operate through different mechanisms and have a more complex impact on the climate.

The net impact of all ABC aerosols is actually to cool the atmosphere so that while it is urgent to reduce BC emissions, care must be taken not to reduce overall emissions in a away that could cause a net detrimental effect. The complexity of the effects of aerosols mean it is vital to carry out research concurrently with taking action to reduce emissions.

Black carbon is attributable to local sources

Worldwide, the major contributors of BC are diesel engines and generators, 'open burning' in agriculture, and domestic biomass burning:

- 42% Open biomass burning (forest and savanna burning)
- 18% Residential biofuel burned with traditional technologies
- 14% Diesel engines for transportation
- 10% Diesel engines for industrial use
- 10% Industrial processes and power generation, usually from smaller boilers
- 6% Residential coal burned with traditional technologies.

However, the sources of BC in Asia are likely to be different to the global average.

While most GHGs are relatively long-lived chemicals (e.g. CO_2), BC and ABC aerosols last just days to weeks. Hence, while other GHGs will travel far from their sources, and so have a global effect, BC concentrations will be peaked at the source. Such hotspots include the Indo-Gangetic plains in South Asia; eastern China; most of Southeast Asia including Indonesia.

As an example of the contribution of one regional source, one study estimates the im-

pact of replacing biofuel cooking with black carbon-free cookers (solar, bio, and natural gas) in South and East Asia would be to reduce BC heating by 70 to 80% over South Asia and by 20 to 40% in East Asia.

It is clear, then that regional emissions have a large impact on regional warming, and the reduction in such emissions would be very important in buying time to reduce the impacts of global warming on the Himalayas. A regional response is needed.

Urban heat islands

Different landscapes absorb solar radiation to different extents. The tarmac, concrete and the shape of the buildings in a city mean that an urban landscape is particularly effective at absorbing solar radiation, and storing the energy absorbed. This leads to higher solar heating in cities than elsewhere, and this along with heat produced by use of energy in buildings and transport means air in cities is hotter than in the countryside. This is called the "urban heat island" effect. It can raise air temperature in a city by 2-8°F.

Evidence shows that the contribution of urban heat islands to warming on a larger scale is minimal – cities occupy a relatively small portion of all land and so the average effect is small. However there is evidence that urban heat islands can have an impact on the climate of immediately contiguous areas, especially those downwind of the city.

The warm air above an urban heat island can be blown by the wind, resulting in a region of warm air known as an "urban plume". It is possible that cities may be heating the Himalayan glaciers, hence contributing to the rate of melting, by this mechanism.

The heat island also affects precipitation. Studies have shown that rate of precipitation is significantly higher downwind of a city than elsewhere – e.g. 28% higher in areas 30-60 km downwind of a selection of American cities.

Measures that can be taken to reduce the heat island effect include planting more vegetation to create shade at ground level and on building roofs, using building and paving materials that are more reflective and water-permeable, taking energy-saving measures to reduce heat produced by buildings, and designing cities such that buildings don't block the wind hence increasing ventilation.

Though urban heat islands are probably contributing little to the regional heating of the Himalayas, they may be influencing the melting of the glaciers and so further research should be undertaken in order to better understand this effect.

Afforestation

In addition to reducing BC and other non-CO_2 GHG emissions, it would be vital to focus on improving the forest cover in the Himalayas to aid the process of mitigating climate change. The Himalayan nations could form a network or alliance of Himalayan forests, similar to the existing network on Boreal forests. This would facilitate research, knowledge sharing and project implementation. Pilot REDD projects need to be launched in each of the Himalayan countries. There is also a need to

provide better resources to foresters in the Himalayan countries to prevent and fight forest fires.

'Contact' effects

The effect of contact with glaciers may contribute to:

- the melt, where 'contact' here refers to such activities as heat-emitting human activity on or very close to the glacier, disposal of waste onto the glacier surface, and building on or next to the glacier. Cases which have been referred to as particularly important have been the military presence on the Siachen glacier and the pilgrim and tourist presence near to the Gangotri glacier, both of which have been claimed to cause significant increased melting. However, the evidence for both of these is contentious and gives rise to ongoing debate, and so while there is no doubt that these activities are detrimental to local ecosystems and the water quality downstream of the glaciers, and hence should be changed, further information should be acquired as to the effectiveness of intervention in reducing regional warming and glacial melt.

Pilgrims and tourists

The Nehnar glacier, situated at a height of 3700m to 4200m around Baltal in northern Kashmir is passed by pilgrims to Amarnath cave. It is thought that heavy pilgrim traffic besides mountain expeditions results in depletion of glacier and environmental degradation. The Indian Expert Committee on Glaciers has recommended the government restricted the number of visitors to these areas.

■ Clearly there is debate as to the actual impact of the military and tourist activity. However, it is clear that the large-scale transportation will emit greenhouse gases, and also BC which will contribute to the global warming. This in addition to the severe pollution caused by the military and tourists must be addressed.

IUCN and the Himalayas
The IUCN Asia Regional Office has been involved in this initiative since early 2009 and has held discussions in China, India, Nepal, and Pakistan and has raised the issue with Bangladesh and Bhutan. To date, the dialogue has been on a Track 2 basis – not formally with governments directly, but with scientific institutes, NGOs, conservation agencies, interested parties, etc.

The most important issue that must be tackled is the emission of BC by regional sources.

While the exact details of the impact of the BC on the glaciers are not completely agreed upon by scientists, it is certain that it causes regional warming over low-lying areas and it is very likely that it also causes glacial melt. It causes changes across the whole region, and of all the factors considered here causes the most intense heating apart from the long-lived GHGs (CO_2) associated with global warming.

International agreements such as the Kyoto Protocol and the Bali Declaration call for achieving desired level of reduction in emissions by 2050. The Himalayas and other similar eco-regions may not have that much time. There is an urgent need for the Himalayan countries [Bhutan, China, India, Nepal and Pakistan] to formulate and execute a joint strategy for the Himalayas before it is too late. The involvement and cooperation of concerned International Agencies is vital.

Opportunities and Policy Implications

- As BC particles are short lived, successful emissions reduction will curb its warming effect in the Himalayas within weeks
- Effective BC emissions reduction may delay the tipping points of major climatic disasters at both regional and global scale, and provide time to develop and implement effective steps for reducing CO_2 emissions
- As soot (and BC) emissions come primarily from the developing nations like China and India in contrast to the industrialized nations in North America and Europe, future burden for emissions control might shift to developing nations
- Technology exists for drastic worldwide reduction in BC

Reduction of black carbon emissions is the most important factor to tackle, because of BC's intense and region-wide impact. Moreover, the short lifetime of emissions means an immediate reduction in heating will result in emissions reduction, which makes it a useful tool in delaying the effects of longer-lived GHGs, thus buying time to mitigate climate change.

There are also additional benefits for tackling BC:

- BC is a constituent of indoor and outdoor particulate matter – a major air pollutant. About 1 million deaths, mostly in developing countries, are attributed to indoor and outdoor air pollution and a range of respiratory infections, cancer and heart disease.
- Unlike CO_2, technology is already available to reduce BC emissions. Examples include, more efficient stoves; reducing biomass burning; installing particle traps on diesel engines; and transitioning to other fuels;
- Reducing the negative effects that BC and its associated atmospheric brown cloud has on the summer monsoon on the Indian subcontinent.

Current evidence is sufficient for us to make a start on reducing non-CO_2 GHGs now; if we need further research this can be done concurrently. This would be a new dimension for tackling climate change. Although we are calling for regional cooperation on this for the Himalayas, the response we propose can also be applied to similar eco-regions elsewhere (e.g. the Alps).

About this note

This note draws largely on the ideas in the paper Stabilisation of Climate Change in the Himalayas by CP Muthanna, Environment and Health Foundation [India] and a subsequent paper The Case for Regional Action to Mitigate Climate Change in the Himalayas by Joseph Schutz, Exeter College, Oxford Uni- versity, UK.

Peter Neil
Climate Change Focal Point, IUCN Asia

IUCN Asia Regional Office
63 Sukhumvit Soi 39
Wattana, Bangkok 10110
Thailand

Tel: + 66 2 662 4029
Fax: +66 2 662 4387
Email: peter.neil@iucn.org

www.iucn.org/asia

International Union for Conservation of Nature

GARUDA RESEARCH & DEVELOPMENT

631/A, Hyder Ali Road, Nazarbad, Mysore – 570010, Karnataka
www.somender-singh.com e-mail : garudarad@gmail.com
Mobile : 98455 36780   Tel 0821 2449018

To the Concerned R&D's / Manufacturers & Think Tanks

Subject : Innovations to improve Diesel Engines and prevent burnouts in Diesel Engines converted to run on CNG / CH4, with extended life.

Given a chance, I can Display & Demonstrate my Patented Innovations related to Refining Internal Combustion Engines with my US Patents : US Patent **US6237579** : Design To Improve Turbulence In Combustion Chambers And my second US Patent **US10393063** : Internal Combustion Engine Piston With Chamber – These Designs relate to Combustion Chambers. The Theory behind my research relates to improving Air & Fuel mixing before and during Combustion which IC Engines lack. Better Mix of Air & Fuel nearing TDC results in Laminar & Controlled Combustion Efficiency with the least build up of emissions in SI & CI Engines.

My decades of Self-funded R&D are based on actual **Cut & Try** methods to achieve more **Km Per Drop Of Fuel** incorporates unique grooves or channels or passages through the squish areas in Combustion Chambers to induce directional turbulence on the compression stroke followed by laminar combustion due to better mixing in the charge on the power stroke.

Back Ground : I have been deeply involved in Motorsport / Racing & Rallying since 1969 in various disciplines with great success. My expertise's in modifying various types of engines resulted in conceiving my 2 US Patents and other innovations. This knowledge comes from modifying New Generation **CRDI & MPFI** Engines at my Tune Up Center @ Mysuru to keep pace with technology by achieving more **Torque & Drivability** with improved reliability with the least carbon formations & accumulation of muck in manifolds as seen in Std New Gen Engines.

Experience gained over decades in improving every possible type of Big & Small Engines gives me the confidence and expertise to improve present **CNG & Diesel** Engines to achieve better Pulling Power in today's urban traffic conditions with lower emissions than what most claim in Dyno Sheets. Recently I did up 2 Tata Cummins CNG Buses in ND resulting in better Torque.

It is an accepted fact that the best **Kmpl** is achieved in the highest gear where the rotation of the wheels are more at any given engine RPM. To achieve this, Torque is far more important than BHP. My tests are based on acceleration from idle speed to full blast in 3 or 4th gears depending on loads without using the clutch by recording the time with a stop clock to measure the gains or losses from 10 Kmph to peak revs / Kmph before the engine cuts out due to rev limiters. After each modification we are sure footed of our work and the rest is left to the owners & operators to give us feed backs on Kmpl and other parameters related to wear & tear, consistency & reliability with driver & passenger comforts due lower vibrations on the long run.

What we have achieved are, better Drivability based on feed backs from experienced drivers & owners and the outcome in fuel savings on highways and urban conditions requiring fewer gear changes due to the increase in usable Torque with reduced Noise & Cabin Vibrations.

My research and experiments show, Air & Fuel Mixing into an ideal charge before ignition play a very vital role in the outcome of efficient combustion which produces the best BMEP to achieve maximum Torque over the widest possible range of operations with least heat into the system.

In practice, to achieve this Ideal state of Air & Fuel Mix in the compressing charge, many forms of turbulence enhancing designs & layouts have been tried including Turbo's and water injection. Injecting fuel into the compressed charge nearing TDC has been the most critical aspect of **DI** Diesels & **GDI :** Gasoline Direct Injection engines to achieve optimum power.
The time duration nearing TDC to achieve proper mixing in the stagnant charge locked inside the bowl at 10 to 20 Bars play a vital role, further affected by ignition delay before Combustion actually starts to occur. As the duration of injection and subsequent combustion play a vital role. <u>Air & Fuel Mixing remains the biggest challenging factors to attain complete combustion in all types of IC Engines, resulting in emissions of UHC, CO, NOx & O3 with less CO2 Per Km / Ton.</u>

In the case of CNG or LNG Gas, the presence of Oxygen under compression also requires proper mixing of Air & CNG along with the spread of Nitrogen which happens to be the largest component / gas present in the compressed gases inside the cylinder to absorb the generated heat out of combustion into rapid expansions to prevent the least heat absorptions into the walls of the combustion chamber / cylinder to achieve maximum Thermal Efficiency with improved BMEP & Torque resulting in the best BSFC. To preventing overheating and oxidation of engine components, Turbulence & Ignition timing in the compressed charge have a direct bearing on Flame Velocity which decides the outcome of Combustion that exerts the maximum pressures on the Piston at Ideal Crank Angles to produce and achieve Max Torque in any IC Engine.

Most in the Diesel World have taken the path to **CRDI : U**sing very high pressure injection exceeding 2000 Bars / 30,000 PSI in the hope of pulverizing the compressed air locked in the combustion chamber with multi squirt fuel injection to produce Max Power with least emissions.

There is very little research or data to show what really happens to Air as compression increases beyond 30 Bars or Hydro Carbon fuels get compressed beyond 1000 Bars. Air is a natural mixture made up of Aprox 3/4 Nitrogen and the rest 1/4 consists of Oxygen and other gases into a natural homogeneous mixture in One Atmosphere on which all Living Life has evolved & existed ever since Life stabilized on Mother Earth. Today **CO2 levels** have gone beyond all known limits in the last Million Years – Based on Carbon Dating inside Polar Ice !

Earthly Fossil Fuels have their own bonding of Hydrocarbons which split, once ignited or when flash points are attained in the presence of Oxygen releasing uncontrolled thermal energy out of a chemical reaction called Oxidization leading to Combustion. It is an accepted fact that no two Cycles of Combustion are the same as you cannot predict Flame Fronts spreading or erupting into spikes during uncontrolled combustion taking place, cycle after cycle in IC Engines.

The only known process to control erratic Runaway Combustion is by introducing Exhaust Gas Recirculation: **EGR** Rich in Carbon Dioxide to quell & suppress erratic uncontrolled combustion preventing Pinging or Detonations. This leads to formations of **CO, HC & O3,** and also resulting in excessive Carbon & Soot formations in inlet manifolds of modern engines.

These are my R&D findings over decades on most types of 2 & 4 Cycle Engines running on fossil fuels in today's World. A google search for – **Somender Singh Grooves** will highlight my work on IC Engines with 100's of Images along with my website : www.somender-singh.com Very few have achieved all this single handed. Further offering Workable Solutions to Reduce Pollution in our congested Urban World's leading to Health related issues in the Young & Old across the World other than issues related to Global Warming.

As a reminder IC Engines consume over 10,000 Liters of Air to burn up one Liter of Fuel producing over 2.5 Kg's of CO2. Aprox 11 Million Barrels of Crude Oils are consumed each day across the World for our mobility. To produce1 Kg of Cement 900Grams of CO2 is the outcome.

-For more details please visit www.somender-singh.com

Somender Singh,
Mysore

MEMORANDUM OF UNDERSTANDING

Between

IUCN, International Union for Conservation of Nature and Natural Resources, a quasi-governmental international organization composed of sovereign States, government agencies, international agencies, non-governmental organizations, and affiliates, founded in 1948 under the auspices of the United Nations Educational, Scientific and Cultural Organization, with its World Headquarters located at Rue Mauverney 28, 1196 Gland, Switzerland (hereafter "IUCN"),

And

The **ASIAN INSTITUTE OF TECHNOLOGY**, an autonomous institute of higher learning established by Royal Charter in the Kingdom of Thailand. The institute is regional in character, its principal place of business being at Km. 42, Phaholyothin Highway, Klong Luang, Pathum Thani, 12120, Thailand ((hereafter "AIT").

Herein referred to jointly and severally as the "Parties" and "Party", as the context may reasonably indicate or require.

Preamble

WHEREAS, the mission of IUCN is to influence, encourage and assist societies throughout the world to conserve the integrity and diversity of nature and to ensure that any use of natural resources is equitable and ecologically sustainable;

WHEREAS, IUCN Asia Regional Office, located at 63 Sukhumvit Road Soi 39, Klongton-Nua, Wattana, Bangkok, Thailand 10110, shall be responsible for execution or implementation of activities agreed between the Parties under the framework of this Memorandum of Understanding ("MoU");

WHEREAS, AIT's mission is to develop highly qualified and committed professionals who will play a leading role in the sustainable development of the region and its integration into the global economy. AIT is based in Thailand and has affiliated centers in other parts of the world;

WHEREAS, Regional Resource Centre for Asia and the Pacific (RRC.AP) – an institute-wide centre of AIT is responsible for execution or implementation of activities agreed between the Parties under the framework of this MoU;

WHEREAS, it is well recognized, based on available scientific evidence, that climate change is causing adverse effects on the Himalayas and the Mekong Region. Further, it is recognized that the climate change effects of the short-lived climate pollutants (SLCPs), in particular of black carbon ("BC") particles, are quite significant, and therefore emission reduction of BC particles would be beneficial in order to avoid further changes in the climate of the Himalayas and Mekong Region, and promoting forest landscape restoration would increase carbon sequestration;

WHEREAS, the Parties share concerns about the effects of climate change on the Himalayas and Mekong Region and recognize the need to promote regional efforts on climate change mitigation and adaptation in the Himalayas and Mekong region, with special focus on emission reduction of BC particles and forest landscape restoration and address the challenges associated with these efforts;

WHEREAS, the Parties intend to work strategically and collaboratively for the formation and operation of the HIMEK Alliance;

WHEREAS, the HIMEK Alliance is intended as a regional cooperation framework between the Himalayan countries and the countries of the Mekong Region to promote efforts on climate change mitigation and adaptation in the Himalayas and the Mekong region to reduce and avoid the adverse effects of the climate change on the Himalayas and the Mekong region, with special focus on emission reduction of Black Carbon particles and forest landscape restoration.

WHEREAS, IUCN and AIT wish to identify projects and activities for joint implementation from time to time through specific agreements ("Supplemental Agreements") for implementing the HIMEK Alliance, and to that end they wish to establish the general terms of their collaboration in this MoU.

Now therefore the Parties agree as follows:

A.   **Objective**

1. The objective of this MoU is to provide a framework of cooperation between the Parties for mutually beneficial cooperation in developing and establishing the HIMEK Alliance among the Himalayan countries and the countries of the Mekong Basin to promote efforts on climate change mitigation and adaptation in the Himalayas and the Mekong region in order to reduce and avoid the adverse effects of climate change on the Himalayas and the Mekong region, with special focus on emission reduction of BC particles and forest landscape restoration (the "Objective").

B.   **Principles of Collaboration**

1.   <u>Complementarity and reciprocal support</u>

The Parties will support each other in working toward the achievement of the Objective and the fulfillment of their respective missions, by building on elements of their respective programs and by pursuing effectiveness while avoiding unnecessary duplication of effort.

2.   <u>Mutual benefit</u>

The specific projects and activities on which the Parties will collaborate under this MoU and through related Supplemental Agreements will be selected, agreed and carried out so as to achieve the Objective and to bring a clear benefit to both Parties and their respective constituents.

3.   <u>Responsibility and funding for collaboration projects and activities</u>

Such collaborative projects and activities will be undertaken with a clear, mutual understanding of the work and the responsibilities to be carried out by each Party and of the ways and means of funding each such project or activity. To that end, the specific details in respect of (a) the work and responsibilities of each Party in terms of operational, performance and administrative tasks, (b) the agreed deliverables to be produced by each Party, and (c) the source, allocation, control and use of all necessary funding shall be set forth clearly in the relevant Supplemental Agreement.

4.   <u>Mutual recognition</u>

Public statements and publications by either Party regarding activities undertaken jointly pursuant to this MoU will expressly acknowledge the cooperative relationship between the Parties. Additionally, where either Party intends to use, in publications in any medium, substantial data and/or information (collectively, "Materials") obtained by the other Party, the Party intending to use the materials will:

a. Give the other Party

  i.   reasonable advance notice of such intended use, and

  ii.   the opportunity to edit or otherwise amend the Materials, or to object to and prevent the intended use thereof, provided that any such objection shall be based on reasonable grounds; and

b. Include with the Materials, in a clearly legible font and conspicuous location in the medium, an acknowledgment or source reference to the other Party.

C.   Areas and Activities of Collaboration

1.   This MoU is intended to provide the Parties with a general framework and a guiding tool in identifying and carrying out specific collaborative projects and activities. Specific areas of collaboration, activities and projects will be identified on the basis of geographic, programmatic and/or other relevant criteria and will be agreed in one or more Supplemental Agreements.

2.   Potential areas of collaboration include, but are not limited to, the following:

a) Drafting of a background paper on Himalaya and Mekong Region elucidating environmental and climate change related issues and identification of activities to address those issues;

b) Drafting of framework to establish the HIMEK Alliance for implementation of activities identified in background paper;

c) On mutual understanding, organise a Secretariat for the HIMEK Alliance that will be responsible for overall coordination and implementation of the identified activities of the HIMEK Alliance;

d) Assist the countries of Himalayas and Mekong region in planning and drafting of a strategy for emission reduction of short-lived climate pollutants (SLCPs), in particular black carbon particles, and forest landscape restoration in the regions;

e) Drafting and submitting a joint project proposal on the HIMEK Alliance to funding agencies in order to secure funds for the implementation of the HIMEK Alliance; and

f) Implementation of activities, such as, conducting research and assessment, publication of reports, organizing workshops, seminars and training,

D. Modalities of Performance

1. In order to ensure the harmonious implementation of the Parties' collaboration and the successful achievement of the Objective of this MoU, the Parties undertake actively to support each other in the performance of agreed tasks/activities and to take all reasonable steps to make the most effective use of the collaboration hereunder in furtherance of their respective missions.

2. Taking into account the preceding paragraph 1, the Parties agree to the following practical steps:

   a. An annual meeting at the senior management level will be scheduled and held every calendar year (the "Annual Meeting"), with the date and venue to be agreed by the Parties in each instance.

   b. The Annual Meeting is intended to provide the Parties with the opportunity to review their collaborative relationship and possibly to extend its scope, and in particular to share information, evaluate past and ongoing joint activities, and discuss new areas and activities for further potential collaboration.

   c. Each Party's key personnel and representative of participating organizations, who are actively involved in specific collaboration activities, will seek to meet on a regular basis, and at a minimum once every calendar year, to review specific aspects of their respective work-plans with a view to achieving or improving complementarity.

   d. Representative of other organizations who are not engaged in collaborative activities will be invited and encouraged to identify possible areas of collaboration between the Parties in their respective regions.

e. Each Party will nominate and notify to the other Party a representative who will serve as focal point dedicated specifically to coordinating the overall collaboration under this MoU, and

ii. the specific activities undertaken pursuant to the specific Supplemental Agreement(s), such as, Memorandum of Agreement (MoA), stipulated by the Parties.

F.   Contact Persons

All communication such as notification and request pertaining to activities under this MoU shall be addressed to:

On behalf of IUCN
1. **Mr. Anshuman Saikia**
   Regional Programme Support Coordinator,
   IUCN Asia Regional Office,
   63 Soi Sukhumvit 39,
   Bangkok, Thailand

   Tel.: +66 2 662 4031 (ext 105); Fax +66 2 662 4387
   Email: Anshuman.SAIKIA@iucn.org

On behalf of AIT
1. **Mr. Shawn Kelly**
   Director,
   Office of External Relations
   Asian Institute of Technology
   P.O. Box 4, Klong Luang
   Pathumthani 12120
   Thailand
   Tel.: +66 (0)2 524 6024
   Fax.: +66 (0)2 524 5069
   Email: director-oexr@ait.asia, oexr@ait.asia

2. **Mr. Osamu Mizuno**
   Director,
   Regional Resource Centre for Asia and the Pacific
   3rd Floor, Outreach Building,
   Asian Institute of Technology.
   P.O. Box 4, Klong Luang,
   Pathumthani 12120, Thailand
   Tel.: +66-2-516-2124
   Fax: +66-2-516-2125
   Email: Osamu.Mizuno@rrcap.ait.asia

g.   Each Party will ensure that any changes made to the list of its focal points will be communicated promptly to the other Party.

G.   Miscellaneous Provisions

1.   <u>Supplemental Agreements</u>

a.   Collaboration activities to be carried out pursuant to any Supplemental Agreement will be:

    i.   Subject to the availability of funds and resources;

    ii.   Approved by the appropriate administrative authorities of each Party; and

    iii.   Undertaken in accordance with the Parties' respective established policies and procedures.

b.   The Parties' performance of Supplemental Agreements shall be subject to and in accordance with the terms and conditions provided for in each such Supplemental Agreement.

2.   <u>Financial Provisions</u>

Financial, administrative and reporting provisions relating to any collaboration activities between the Parties shall be expressly agreed in the relevant Supplemental Agreement.

3.   <u>Dispute Resolution</u>

Any dispute arising out of or in connection with this MoU will be settled by amicable negotiation between the Parties. Should the Parties be unable to negotiate an amicable settlement, the dispute shall be submitted to conciliation following procedures to be agreed by the Parties.

4.   <u>Representation</u>

Neither Party shall have the authority to incur any liability or make any commitment on behalf of the other Party *vis-à-vis* any third party, contractually or otherwise, without the other Party's advance express written consent.

5.   <u>Amendment</u>

This MoU may be amended only by a writing signed by both Parties.

6.   <u>Term and Termination</u>

This MoU shall become effective on the date of signature by both Parties, and shall remain in effect for five (5) years from the effective date unless renewed in writing for a similar term or terminated by either Party. Either Party may terminate this MoU by giving the other Party six (6) months' advance written notice of termination. It is understood that any such termination shall have no effect on any Supplemental Agreements then in force between the Parties, and that the performance of such Supplemental Agreements shall be subject to their own terms and conditions.

7. <u>Property Rights</u>

Neither Party shall have the right to use the other Party's name, logo and/or other trademarks in any medium and for whatever purpose without the other Party's prior written consent in each instance of use.

8. <u>Non-enforceability</u>

This MoU is a non-binding statement of the Parties' mutual understanding of their proposed collaboration framework. Therefore, and except for the obligations set forth under Article G7 above, this MoU is not intended to create, and does not create, any legally enforceable rights or obligations in respect of either Party, including any obligation on their part to enter into any Supplemental Agreement.

In witness whereof, the undersigned, being duly authorized to do so, have executed this MoU in the English language in two (2) counterparts, each of which shall be deemed an original, and which together shall constitute one and the same instrument.

IUCN, International Union for Conservation of Nature and Natural Resources

Signed: _______________________

Ms Aban Marker Kabraji,
Regional Director, Asia and Director - Regional Hub for Asia and Oceania

Date: July 4, 2018

Asian Institute of Technology

Signed: _______________________

Prof. Worsak Kanok-Nukulchai
President

2 2 MAY 2018

Date: _______________________

<u>Witnesses</u>

IUCN, International Union for Conservation of Nature and Natural Resources

Signed: _______________________
Dr TP Singh,
Deputy Regional Director, Asia

Date: 4 July 2018

Asian Institute of Technology

Signed: _______________________
Mr. Osamu Mizuno,
Director, AIT Regional Resource Centre for Asia and the Pacific

Date: 23 May 2018

Brick Kilns in India: Environment Challenges & Green Technologies
Dr Sameer Maithel, Director,Greentech Knowledge Solutions Pvt Ltd
January, 2022

Key environmental challenges facing the brick industry

The environmental challenges facing the Indian brick industry can be broadly classified into three categories:

Air Pollution and health

Air pollution is the presence of substances (gases, particulates, and biological molecules) in the atmosphere that are harmful to the health of humans and other living beings, or cause damage to the climate or to materials[1]. There are several sources of air pollution in brick kilns. The combustion or burning of fuel (mostly coal and biomass fuels) for firing the bricks gives rise to air pollution. The pollutants consist of particulate matter of various particle sizes ($PM_{2.5}$, PM_{10}) produced due to incomplete combustion (which gives black colour to the smoke) along with gaseous pollutants, such as, Sulphur dioxide (SO_2), Carbon monoxide (CO), Hydrogen Fluoride (HF), etc. The combustion related pollution is often referred as stack emission or emissions discharged through the chimney or stack. In addition to the stack emission, dust pollution is also caused as the large amount of clay, ash, powdered fuel at the brick making site gets entrapped in the air or dust pollution caused by the movement of trucks, tractor trolleys etc. on the unpaved dusty roads around a brick kiln.

Air pollution is a major environmental health problem and India is one of the most polluted country in the world in terms of concentration of particulate matter ($PM_{2.5}$, PM_{10}) in the air. PM_{10} are inhalable coarse particles, which are particles with a diameter between 2.5 and 10 micrometers (μm) and $PM_{2.5}$ are fine particles with a diameter of 2.5 μm or less. Particulates are the deadliest form of air pollutants due to their ability to penetrate deep into the lungs and blood streams unfiltered. The smaller $PM_{2.5}$ are more harmful as it can penetrate deep into the lungs. As per a recent report, 21 of the world's 30 cities with the worst air pollution are in India [2]. Most of these Indian cities are located in the Indo-Gangetic plains. Vehicles, industries, thermal power plants, construction, biomass burning, diesel gensets, commercial and domestic use of fuel, are identified as major sources of air pollution.

Emission of CO_2 and global warming

Greenhouse effect, a warming of Earth's surface and troposphere (the lowest layer of the atmosphere) caused by the presence of carbon dioxide, water vapour, methane, and certain other gases in the air[3]. Global warming is leading to changes in the earth climate which turn have impacts on the rainfall patterns, agriculture, forests, biodiversity, human health, etc. The main reason for global warming is the increasing concentration of carbon dioxide in the earth's atmosphere due to the burning of fossil fuels (coal, petroleum fuels). Fossil fuels contain carbon which on combustion gives rise to carbon dioxide. Coal is the main fuel in brick kilns in India and the brick industry in India is estimated to contribute 66–84 million tonnes of CO_2 emissions per year[4], which is about 3% of the total CO_2 emissions of the

---

[1] https://en.wikipedia.org/wiki/Air_pollution
[2] https://edition.cnn.com/2020/02/25/health/most-polluted-cities-india-pakistan-intl-hnk/index.html
[3] https://www.britannica.com/science/greenhouse-effect
[4] TERI. 2016 Report on Resource Audit of Brick Kilns New Delhi: The Energy and Resources Institute [Project Report No. 2015IE22]

country.

Removal of clay from agriculture fields and its potential impact on agriculture

Clay is generated from the decomposition of rocks and is one of the most abundant natural mineral materials on earth. It is the primary raw material for manufacturing bricks. The clay used for brick manufacturing must have certain properties like plasticity, which permits it to be moulded when mixed with water. It should also have the ability to retain the shape while drying and desired vitrification properties for proper firing. The brick making clays are either surface clays or shales. Surface clays are found near to the surface of the earth. On the other hand, shales are clays that have been subjected to high pressures under the earth surface and must be mined. In India, brick production depends on surface clays e.g. clay obtained from agriculture fields, surface clay washed into water bodies (tanks, dams, etc.) and harnessed by desilting or clay harnessed from rivers in the river delta regions. For producing 250 billion solid bricks every year in India, approximately 750 million tons[5] of brick earth or clay is required and a significant part of it comes from agriculture fields and there are concerns related with loss in agricultural productivity and degradation due to unplanned mining.

Green Technology Options & Potential

A list of green technology options is discussed in the following paragraphs.

Converting FCBTKs to Zig-Zag Kiln and other Cleaner Brick Kiln Technologies

Zig-zag kiln in the form of High Draught kiln was introduced in India around 50 years ago by the Central Building Research Institute (CBRI) as an alternative to the Bull's Trench Kiln. The main difference in the two kilns is regarding the brick setting and air flow. In a zig-zag kiln the bricks for firing are set/loaded in such a way that air in the kiln follows a zig-zag path, while in a BTK the air follows a straight line path. The introduction of a fan in the original design of the High Draught kiln also provided more air for combustion. The coal is crushed/powdered and is fed in small quantities. All of these changes help in more efficient heat transfer and better burning of the fuel. A zig-zag kiln saves around 20-25% fuel and results in up to 50% reduction in particulate matter emission as well as improvements in the percentage of good quality bricks.

---

[5] Assuming 3 kg of brick earth is required for the manufacturing of 1 solid clay brick.

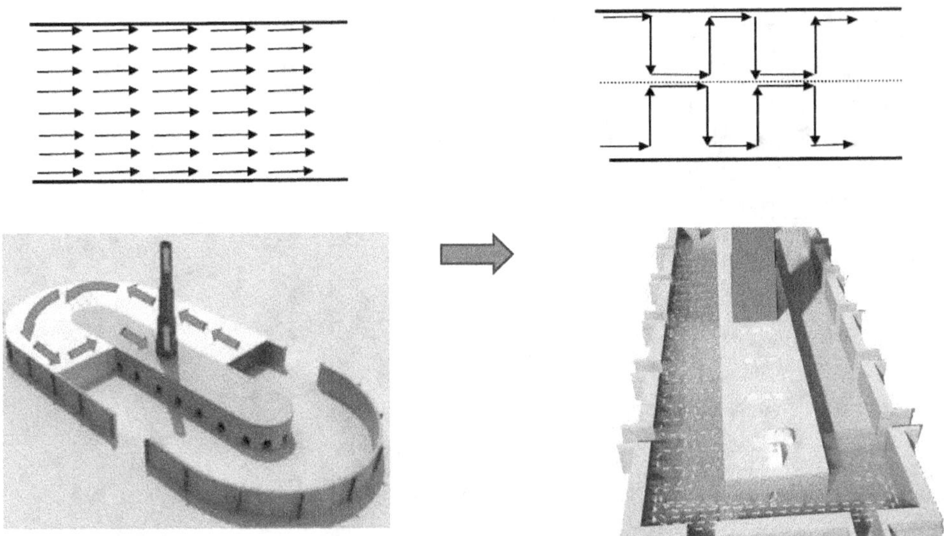

Figure 1 Conversion of FCBTK into zig-zag kiln

During last 10 years, the technology has seen rapid adoption, mainly driven by the mandatory enforcement by State Pollution Control Boards, in the states/regions of Punjab, Haryana, districts of West Uttar Pradesh falling under the National Capital Region (NCR), Bihar and some districts of East Uttar Pradesh. During this period, the technology has got introduced in Tripura, Rajasthan, Maharashtra, Odisha, etc. It is estimated that now (January 2022) India has around 15,000 zigzag kilns. The cost of retrofitting an existing FCBTK to zig-zag kiln can cost Rs 10-30 lakhs depending on extent of retrofitting required. The initial investment for constructing a new good quality zig-zag kiln is estimated to be around Rs 40 lakhs.

Among brick makers there is a consensus about the multiple benefits of the technology. Conversion of FCBTK to zig-zag kiln is a beneficial proposition for the entrepreneur as it can lead to higher profits because of savings of fuel as well as additional revenue due to the increase in the percentage of good quality bricks. There are potentially around 40,000 existing FCBTKs which are yet to be converted to zig-zag kiln. The largest number of unconverted kilns is in the state of Uttar Pradesh. Taking a conservative estimate of savings of 100-150 tons of coal per year per kiln, if all the remaining kilns are converted that would result in savings of around 5 to 7.5 million tons of coal every year. It is expected that mandatory regulations by State Pollution Control Boards will force these conversions over next five years or so. However, there is a strong case for coordinated awareness generation among brick makers, making available standard designs and trained manpower for both construction and operations as well as providing access to finance to smaller brick makers so that they are able to adopt the technology.

Box 1: Converting FCBTKs to zig-zag kiln technology

Environmental Benefits:

- Reduction in coal consumption (20-25%) and associated CO_2 emissions
- Reduction in emission of black carbon and particulate matter (up to 50%)
- Reduction in wastage and improved quality

Investment requirement

- Rs 10 lakhs -40 lakhs/ kiln

Status & Potential

- Adopted by around 15,000 kilns till January, 2022
- Around 40,000 FCBTKs yet to adopt; can result in savings of 5 to 7.5 million tons of coal/year

Action Points

- Intense awareness campaign amongst brick enterprises
- Making available standard designs
- Skill training of kiln masons
- Skill training of supervisors and workers involved in brick setting and fuel feeding
- Financing arrangements for smaller kilns

Apart from the conversion to zig-zag kiln technology, the other options for reducing air pollution and improving the brick quality are conversion of existing FCBTKs to Hoffmann kiln or Tunnel kiln technologies[6]..

- Tunnel kiln is a continuous moving ware kiln in which the clay products to be fired are passed on cars through a long horizontal tunnel. The firing of products occurs at the central part of the tunnel. The tunnel kiln is considered to be the most advanced brick making technology. The main advantages of tunnel kiln technology lie in its ability to fire a wide variety of clay products, better control over the firing process and high quality of the products. The typical construction cost of a tunnel kiln ranges from Rs 5 -10 crores. In India, there are very few (around 10) tunnel kilns operational for brick manufacturing.

- Hoffman kiln is a continuous, moving fire kiln in which the fire is always burning and moving forward through the bricks stacked in the circular, elliptical or rectangular shaped closed circuit with arched roof. The fire movement is caused by the draught provided by a chimney or a fan. The typical construction cost of a Hoffmann kiln and its new variant Hybrid Hoffmann kiln ranges from Rs 1.5 -5 crores. Hoffmann kilns have been used in India, primarily to produce clay roofing tiles, and more than 100 Hoffmann kilns are estimated to be operational in Kerala, Karnataka and Balaghat (Madhya Pradesh).

Both these options (Tunnel kiln and Hoffmann kiln) require a much larger capital investment. If these technologies are used only for firing common solid bricks the cost of production is higher and it will be difficult for the brick enterprises using these technologies to compete with enterprises having zig-zag kiln technology. The viability of these kiln technologies improves if they are primarily used for the production of value-added products like hollow and perforated bricks, clay tiles, etc.

---

[6] The fact sheet on Tunnel kiln and Hoffmann kiln can be downloaded from https://www.gkspl.in/wp-content/uploads/2018/10/REBM3.pdf

Figure 2 Tunnel Kiln[7]

Mixing Fly Ash, other industrial wastes and internal fuel with Clay

Addition of fly ash and other industrial wastes like boiler ash in clay for making bricks is widespread in central, western and parts of southern India. The mixing is done as a part of clay mix preparation before moulding of bricks. The addition of fly ash and wastes with clay helps in reducing the amount of clay required for making bricks. In addition, addition of fly ash which also has some unburnt carbon in it also helps in reducing the amount of fuel required for baking the bricks. Addition of fly ash in black cotton soil (found widely in central and western India) is also beneficial in reducing the plasticity of the soil and hence reducing breakages taking place during drying and firing. In addition to fly ash, internal fuel in the form of powdered coal, rice husk, bagasse, boiler ash, etc is also added with clay. Use of internal fuels can also help in reducing air pollution from brick kilns. It can also help in reducing the density of bricks and thus improving the thermal insulation provided by them.

There are many ways in which mixing of fly ash or waste with clay is practised:
- Manual mixing (Figure 4)
- Mixing using a tractor attachment (Figure 4)
- Mixing using a mechanical mixer/pug mill

Depending upon the technology employed the investment can range from a few lakhs with maximum going up to Rs 20 lakhs.

.

---

[7] https://www.gkspl.in/wp-content/uploads/2018/10/REBM3.pdf

Figure 3 Mixing of fly ash with clay (Maharashtra & Telangana)

With multiple environmental benefits, there is a large potential for scaling-up the mixing of fly ash, industrial wastes and internal fuel with clay, particularly in the Indo-Gangetic region, where use of internal fuel can also help in reducing air pollution. It is to be noted that the amount of fly ash, industrial waste and internal fuel which can be mixed with clay, depends on the properties of the clay. For example, while it is possible to mix larger quantities of fly ash in clays in central and western India, only smaller amount of fly ash can be mixed in clays in northern and eastern India. An organised initiative aimed at mapping the available industrial wastes, potential internal fuel and working out feasible clay-waste-internal fuel mixes is needed. Some improvements and modifications in the mixing technologies and machinery may also be required depending on the type of waste and the scale of production.

Box 2: Mixing Fly Ash, other industrial wastes and internal fuel with Clay

Environmental Benefits:
- Utilisation of industrial and other wastes
- Reduction in clay consumption
- Reduction in fuel consumption
- Reduction in emission of particulate matter in stack gases
- Potential to produce porous bricks having better insulation properties

Investment requirement
- Up to Rs 20 lakhs/ kiln

Status & Potential
- Common practice in parts of Central, Western & Southern India
- Potential to expand to around 60,000 FCBTKs/ zig-zag kilns located in the Indo-Gangetic plains

Action Points
- Testing of clays
- Mapping the availability of wastes and internal fuel
- Demonstrating viable clay-waste-internal fuel mix
- Adapting mixing technology as per the requirements of the Indo-Gangetic plains
- Skill training of supervisors and workers involved in brick setting and fuel feeding
- Financing arrangements for smaller kilns

Use of mechanical coal feeding system in zig-zag and Hoffmann kilns

The coal or fuel feeding in existing FCBTK, zig-zag and Hoffmann kilns is carried manually by firemen. Usually, they take coal in a spoon and feed it in the feeding holes. At a time, coal is fed by one or two firemen. Usually, the feeding of coal is not continuous, coal is fed for around 10 minutes followed by a non- feeding interval of 15-20 minutes after which again the coal/fuel feeding is repeated. This gives rise to black carbon and particulate matter emissions. A continuous feeding of properly sized fuel in a zig-zag kiln will reduce both coal consumption as well as emissions by ensuring cleaner combustion. Various types of mechanized coal stoking systems or solid fuel burners can be employed for the purpose. The solid fuel burner of a particular type is shown in Figure 5 includes a fuel crushing and distribution unit. It delivers solid fuel mixed with positive airflow ensuring perfect and consistent firing.

The use of mechanical coal feeding systems will not only reduce the air pollution from brick kilns but can also improve immensely the working conditions of the fire men. Currently the firemen walk on the kiln surface (surface temperature 80-120 °C) for feeding fuel and hence exposed to excessive heat. Zig-zag and FCBTK offer some specific challenges for the use of feeders as they do not have permanent roof, a technical solution would have to be found for the placement and movement of the feeder system.

Box 3: Use of mechanical coal feeding system in zig-zag and Hoffmann kilns

Environmental Benefits:
- Reduction in fuel consumption
- Reduction in emission of particulate matter in stack gases
- Improvement in quality of bricks

Investment requirement
- Rs 10 to 40 lakhs/ kiln depending on the type of system

Status & Potential
- Trickle feeder in use on Hoffmann kilns at Balaghat
- Designs available, yet to be tested on zig-zag kilns
- Potential to expand to kilns located in NCR and other heavily polluted regions in the Indo-Gangetic plains

Action Points
- Pilot testing
- Making available affordable technology
- Training of owners and workers in operation

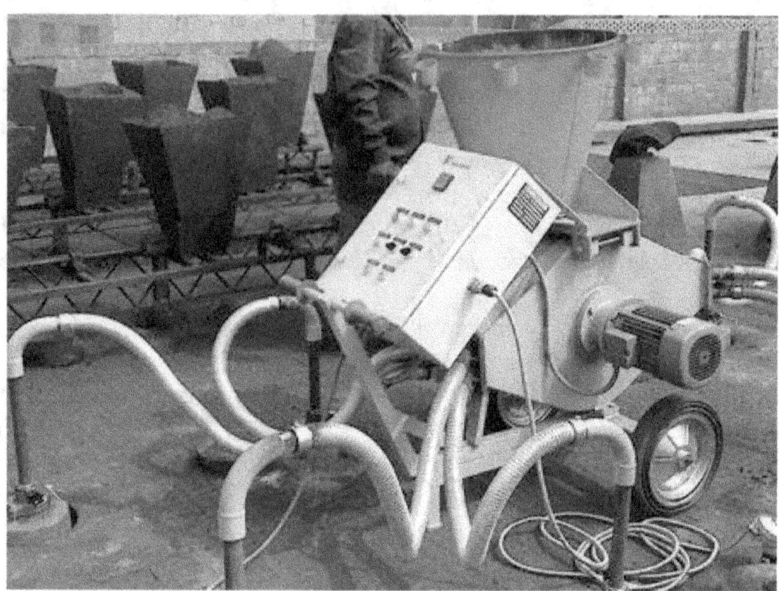

Figure 4 An octopus type coal feeder consisting of a coal crusher and delivery unit [8]

Use of gaseous fuels in brick kilns

Gaseous fuels burn more cleanly and produce less particulate matter emissions compared to solid fuels like coal and biomass. The National Green Tribunal (NGT) in its recent orders has mentioned about exploring possibility of using gaseous fuels in brick kilns located in the NCR region. Gaseous fuels like piped natural gas, compressed natural gas or biogas are cleaner fuels which results in less air pollution as well as they also help in reducing CO_2 emissions.

Recently, Wienerberger India (a large international company involved in manufacturing of bricks), has converted their factory at Kunigal, Karnataka to natural gas firing[9]. In Tripura, three FCBTK brick kilns were using natural gas from 1998-2008[10]. In neighboring Bangladesh, several Hoffmann kilns operated on natural gas (Figure 6). Conversion of zig-zag kilns to gaseous fuels will require development of appropriate technology in the Indian context. Some of the key questions which has come out during discussions with brick industry representatives are:

- Whether it is technically possible to retrofit the existing zigzag kilns for natural gas firing?
- Brick kilns are located amidst agricultural fields often in remote places, how would the gas supply be ensured?

---

[8] Source: http://www.beralmar.com
[9] https://www.wienerberger.in/Press/News0.html
[10] Information shared by Member Secretary, Tripura State Pollution Control Board, during the webinar on "Ensuring Decent Work and Green Technology in India's Brick Kilns: Issues and Challenges" organized by CEC on December 23, 2020

- The price of natural gas is almost three times that of coal that will increase the cost of firing the bricks. In addition, substantial investments would be required by the brick kiln owners in converting their kilns to gaseous fuels. Would the use of gaseous fuels make clay bricks costly and economically non-viable compared to other alternate products like AAC blocks?
- In case of piped natural gas, whether the gas companies would be ready to supply gas to seasonal industries like brick kilns?

If we look at the example of Europe, the brick industry changed over from coal to natural gas over a 20 years period (1960 to 1980)[11]. This period was also characterised by mechanisation and the consolidation of brick industry.

Box 5: Use of gaseous fuels in brick industry

Environmental Benefits:
- Reduction in emission of particulate matter and pollutant gases
- Reduction in energy consumption and CO_2 emission
- Reduction in wastage and improvement in quality

Investment requirement
- Estimates not available

Status & Potential
- Need for pilot testing to carry out techno-economic feasibility
- Could be a longer-term solution for NCR and other heavily polluted regions in the Indo-Gangetic plains

Action Points
- Pilot testing to carry-out techno-economic feasibility
- Understand implications on the industry and to develop a roadmap

Figure 5 Use of Natural Gas in a Hoffmann Kiln in Bangladesh (Source: GKSPL)

---

[11] https://www.brickguru.in/en/blog/moving-to-natural-gas-for-brick-firing-is-it-possible-in-india/

Currently almost the entire brick production in the country is of the solid bricks, with less than 1% being that of hollow and perforated bricks[12]. However, a significant part of the demand for solid bricks can be replaced with that of hollow and perforated bricks. The production of hollow and perforated bricks addresses all major environmental concerns as it can help in reducing the requirement of clay by up to 55%, reduce energy consumption and CO_2 emissions by up to 55 % and can result in very large reductions in air pollution[13]. Also, the perforated and hollow products offer better thermal insulation so would help to reduce operational energy required for cooling or heating a building.

There are two routes to make a shift towards the manufacturing of hollow and perforated bricks:
a) Upgradation of existing zig-zag or Hoffmann kilns: In this case the existing brick kilns can mechanise their clay preparation process and install an extruder to manufacture green bricks having holes/hollows. The extruded bricks require controlled drying and hence drying under a shed or using an artificial dryer is required. The hollow and perforated bricks can be fired in the existing kilns. This upgradation has been operating in around 30 to 50 kilns in the country. The initial investment of upgrading the clay preparation, installing an extruder and putting up a drying shed can range anywhere between Rs 2 to 6 crores. This upgradation can help brick enterprises to enter the perforated and hollow brick market, but without having an artificial dryer the production volume remains small and it is difficult to reduce the cost of production.

b) Setting up a mechanised tunnel kiln and artificial dryer plant for exclusive production of hollow and perforated products: In this case a dedicated plant consisting of clay preparation machinery, deairing extruder, artificial dryer, tunnel kiln and mechanised material handling systems is installed. Such a plant generally has capacity to produce minimum 1 lakh regular bricks per day. In these plants it is possible to bring down the cost of production and maintain consistent quality. The initial investment in plant and machinery is of the order of Rs 15-25 crores.

Shifting to hollow and perforated bricks requires careful testing of the clay as not all clays are suitable for extrusion. Thus, a suitable clay mix must be prepared. For running a mechanised brick plant, unhindered supply of electricity is needed. Shifting to hollow and perforated bricks production means a structured transformation of the brick industry and converting brick industry into an organised manufacturing industry. Such a transformation would require larger investments, access to credit and technology, technically skilled manpower for the operation of machinery, higher technical and managerial capacities, and year-round operations. With year-round operations, brick industry will not be able to work with seasonal migrant workers. This would translate into better paying and more stable jobs for workers and also reduction in drudgery and better working conditions. On the other hand, with a shift from manual to mechanised operations, there would a reduction in the number of workers and hence overall reduction in the employment in brick industry. While a minority of progressive brick makers welcome the idea of such a transformation, majority of brick makers find this very intimidating.

---

[12] GKSPL 2016. Market Assessment for Burnt Clay Resource Efficient Bricks (REBs), Greentech Knowledge Solutions Pvt Ltd, New Delhi
[13] http://www.resourceefficientbricks.org/pdf/REB_booklet_Mar2017.pdf

Figure 6 A Hollow Block

Box 6: Manufacturing of hollow and perforated bricks

Environmental Benefits:

- Reduction in clay usage
- Reduction in emission of particulate matter and pollutant gases
- Reduction in energy consumption and CO_2 emission
- Reduction in wastage
- Better thermal insulation of walls

Investment requirement

- Upgradation of existing facility: Rs 2-6 crore
- New tunnel kiln based mechanised plant: Rs 15-25 crore

Status & Potential

- 30-50 kilns who have upgraded their process to produce at least part of the production as hollow or perforated products
- Very few successful models of dedicated plants
- Medium to long term solution

Action Points

- Demonstration of business models
- Access to finance
- Medium to long term policy certainty, particularly around raw material supply
- Indigenous production of machinery, tunnel kiln and dryer systems

BIBLIOGRAPHY

1. Climate Briefing Note: 9 June 2008; Institute for Governance and   Sustainable Development/International Network for Environmental Compliance and Enforcement

2. G. Carmichael, V. Ramanathan; Nature Geosciences, 2008 Vol 1, Issue 4, **pp 221-227**

3. Gopal Rawat Eric D. Wikramanayake, Pralad Yonzon, Himalayan   subtropical pine forests (IM0301); WWF Report, 2001

4.Kathy S Law, Andreas Stohl, 'Arctic Air Pollution: Origins and   Impacts; Science Magazine March 2007, pp 1537-1540

5. 'Protecting Life in the Ganga' Climate Contours- WWF Report, July,   2007, Page 6

6. Stenlund Peter; 'Lessons in Regional Cooperation from the Arctic,' Ocean and Coastal Management Journal, 2002, vol 45; pp 835-839

7. Climate Impacts and Mitigation Costs of Non CO2 Gases Paper by PEW Centre on Climate Change John M Reilly, Henry D Jacoby, Ronald G Prinn, Massachusetts Institute of Technology

8. Localizing Climate Change: Controlling Greenhouse Gas emissions in the United States Michele M Betsill, Belfer Center for Science and International Studies.

9.Articles regarding formation of Arctic Council, Printed in the Journal 'Northern Perspectives', published by the Canadian Arctic Resources Committee. [Vol 19, No:2, Summer 1991]

10. Climate Change and Air Quality—Measures with Co-Benefits in China

Kristin Aunan, Center for International Climate and Environmental Research-Oslo (CICERO)

11.Patented Vehicle technology by Somender Singh
www. somender singh. com grooves

12. Measures to Mitigate Urban Heat Islands
Yoshika Yamamoto, Environment and Energy Research Unit, Science and Technology Foresight Center, Tokyo

12. NASA Report on Urban Heat Islands

14. Geomorphologic evidences of retreat of the Gangotri glacier and its Characteristics, Ajay K. Naithani*, H. C. Nainwal, K. K. Sati and C. Prasad Department of Geology, HNB Garhwal University.

15.WWF Study on Siachen Glacier
Arshad H Abbasi

16. Report of the Task Force on The Mountain Ecosystems [Environment and Forest Sector] for Eleventh Five Year Plan, Planning Commission, Government of India

17. Report on Ambient Air Quality of Kathmandu Valley [2005] Ministry of Environment, Science and Technology, Kathmandu

18.Review of improved cook-stoves programme in Himachal Pradesh
Dr. Y.S. Parmar University of Horticulture and Forestry, Nauni, Solan HP, India

19. Report of Working Group 2 [WG2] of the IPCC on Climate Change and Air Pollution – a long term perspective

20. Reducing Black Carbon May Be Fastest Strategy for Slowing Climate Change
IGSD/INECE Climate Briefing Note: 9 June 2008

21. Primer on Short Lived Climate Pollutants: Institute for Governance & Sustainable Development -IGSD Working Paper: November 2013

22. Soot, Solidarity and Survival on the Roof of the World; FIRE AND ICE by Jonathan Mingle, 2015.

23. Top climate scientists are skeptical that nations will rein in global warming
Jeff Tollefson Nature Briefing Newsletter 01 November 2021

24. What is green hydrogen? Can India make it affordable?
   Harshit Rakheja, Business Standard, November 25, 2021
